PATENT
SECRETS

Disclaimer

This book contains the experiences and advice of the authors and should in no way be considered legal advice or instructions. The authors are not engaged in rendering any legal service. Therefore, the services of a professional are recommended if legal advice or assistance is needed.

The authors and publisher disclaim any responsibility for personal loss or liabilites caused by the use or misuse of any information presented herein. *This book is for adademic study only.*

PATENT SECRETS

HOW YOU CAN PROTECT YOUR INVENTION FOR AS LITTLE AS $25

STEVEN HAMPTON AND CRAIG HERRINGTON

PALADIN PRESS · BOULDER, COLORADO

Also by Steven Hampton:

Secrets of Lock Picking
Advanced Lock Picking Secrets
Security Systems Simplified

Patent Secrets: How You Can Protect Your Invention for as Little as $25
by Steven Hampton and Craig Herrington

Copyright © 2000 by Steven Hampton and Craig Herrington

ISBN 1-58160-124-7
Printed in the United States of America

Published by Paladin Press, a division of
Paladin Enterprises, Inc.
Gunbarrel Tech Center
7077 Winchester Circle
Boulder, Colorado 80301 USA
+1.303.443.7250

Direct inquiries and/or orders to the above address.

Visit our Web site at www.paladin-press.com

TABLE OF

Contents

Preface

First of all, *there is no bad idea*. There are only bad inventions.

How can an invention be bad? Either it does not address the need of the people, or it is blundered into extinction because of a lack of knowledge on the inventor's part. We have professional experience in the latter. But our loss is your gain: We know where you are coming from and where you could possibly go.

As a lone inventor, you need help from somewhere, and we traveled that road alone for almost 30 years. Want to know how we finally beat the system? What we learned to make millions of dollars? How you can sell your invention *before* you patent it! The most wonderful thing about being an inventor today is that you can become wealthy without jeopardizing your lifestyle or even leaving your hometown! Since you bought this book and you are reading these words, we can safely assume that you are intelligent enough to wonder how protecting your creation for $25 is possible. The fact is, there *are* a few cases where our government has actually made laws that protect you, the little guy or gal. And allowing you the opportunity to protect your idea for a limited time at very little cost is one of them.

But in addition to protecting your invention for a small fee, there are other options available to you. As a budding inventor, your ultimate goal is probably to get your product on the market and make as much money as possible. This book was written to inform you of *all* of the options open to you as you pursue that goal—and the things you should take into consideration before you even begin. Having walked in your shoes, we have information you need right now, but we had to learn it the hard way. Our goal in writing this book was to help you make informed choices—and hopefully, in the process, save you a fair amount of grief and a whole lot of money.

Introduction

When the U.S. Constitution was established, article 1, section 8 provided that "Congress shall have the power . . . to promote the progress of science and the useful arts by securing for limited times to authors and inventors the exclusive right to their respective writings and discoveries."

Shortly thereafter, on April 10, 1790, a three-man Patent Board was established, consisting of the secretaries of state and war and the attorney general.

It wasn't long before inventors started showing up. As secretary of state (and a leading inventor himself), Thomas Jefferson did most of the work, which consisted of listening to each inventor and then granting him a patent neatly signed by George Washington.

Evidently, inventors were just as long-winded in those days as they are now, because by 1793 everyone on the Patent Board, including Jefferson, threw up their hands. It was just too much. A new law was passed that made the issuance of patents a mere matter of filing some papers and paying some fees, after which a clerk gave the inventor his patent.

Having done away with the critical examination of applications, the patent office was handing out patents almost indiscriminately. As a result, such confusion and protest arose over conflicting claims, duplicate patents, patents of old ideas, and so forth, that the whole business was a mess.

In 1836 Congress enacted the first patent law, providing for the establishment of a commission and a corps of examiners and empowering the commissioner to reject or grant applications. This law, although amended from time to time, remained the basic Law of Patents until January 1953. It has since been modified by scores of other acts of Congress and hundreds of court decisions.

These judicial decisions created an immense amount of new *common law* (as distinct from *statute*)—some of it good, some of it bad, and a lot of it contradictory and confusing.

One fundamental question raised in these patent lawsuits was the precise definition of "invention." Just what is it? And where do you draw the line between mechanical skill and genius? Let's take an example.

Suppose you are the owner of a small machine shop, and, at the suggestion of your brother-in-law, you have developed a special machine to crank out jiggle-maroos. He has tramped the streets of the big industrial towns, slept in roach-infested motels, hitchhiked to save train fare, and finally succeeded in getting the machine into the factories of the big jiggle-maroo makers. You have gone without most of the things that make life worth living in order to finance the materials and labor for turning out the machines.

Then it hits the fan. Some character you have never heard of suddenly shows up at your door with a patent and a slick lawyer, informing you that you are infringing on his "rights" and demanding retribution. According to him, this entails your giving him 90 percent of your profits and handing over control of your company or going out of business. You seem to be confronted with two equally unappealing alternatives: either go out of business entirely or surrender and become a slave in the plant where you were once master (this same situation can—and very often does—occur whenever you take in investors). Or is there some other way to wiggle out?

You ask your own lawyer, and you find that there may be. Did you and your brother-in-law apply for a patent on your machine, he asks. Of course not, you reply. You didn't think it was an invention. It was just a good idea. Anybody could have thought it up if he was familiar with the jiggle-maroo business. It didn't take any particular brains. What the hell did the government mean by giving this guy a patent on a machine that any skilled mechanic could create?

"Aha!" exclaims your lawyer. "Tell the patentee to sue. Our defense will be that his patent is invalid for lack of invention. We will summon a battery of expert witnesses from all the leading jiggle-maroo factories and convince the court that the patent office goofed it up."

So the two of you slug it out in court. You maintain that the machine is merely a product of mechanical skill and cannot be dignified with the word "invention." Your opponent says that it is obviously more than mechanical skill, that he was the first to conceive of the idea, and that the patent office recognized this fact and thought it was definitely "invention" and worthy of patent protection.[1]

The above scenario seldom occurs nowadays. The courts are backed up for years, and by the time you are pulled into court, you may have already made the full profit potential or lost your butt. In either case, the fellow with the slick attorney will be the one who loses. After several, or even a dozen, years, the judge may rule in his favor; however, you can always appeal. Furthermore, award collection is not enforced by the Patent Office or the courts; the inventor is responsible for enforcing his own award. In our high-paced society, driven by a world marketplace, the first one to market ultimately wins.

As an inventor, you have three major avenues to wealth and success, and all of them involve protecting your idea in one way or another. In the following pages, we will try to help you determine for yourself the one that will allow you to make the most possible money from your idea with the least amount of legal hassle.

NOTES

1. Alan Montague, "What the Patent Law Means to the Inventor," *Science and Mechanics Magazine* (September 1962): 15.

CHAPTER 1

What Are Your Options?

Basically, when charting your course for developing your invention, there are three major routes you can take. You can (1) get a patent and seek investors, (2) file a disclosure document (Form PS-1b, which provides for a two-year limited protection of an inventor's idea for a small fee) with the U.S. Patent and Trademark Office (PTO) and then license your product to a second or third party who will market it for you, or (3) protect your idea yourself and go straight to market.

The advantage of obtaining a patent, of course, is legal ownership, which gives you the moral high ground in the eyes of the law—"protection" for a 20-year period. (Although you should consider the fact that your protection is unenforceable and expensive from a practical point of view should you have to go to court to defend your patent rights.)

On the downside, obtaining a patent necessitates the lengthy and complicated process of finding a good attorney, having his associates do a prolonged patent and claim search, drawing up the legal documents, filing, and waiting. Patenting ideas for the sake of owning patents is a rather expensive pastime. For a 1999 utility patent, the basic filing fee, issue fee, and three maintenance fees required during the patent term totaled approximately $4,000 for a small entity (any company under 500 people). If you fail to pay your maintenance fees, you lose all rights to the invention. And you have yet to develop and sell your invention! Most inventors who are awarded a patent will then look for investors.

The advantage of filing a disclosure document (see Appendix II) is that you have legal proof of your concept on file with the PTO before trying to sell your idea to a partner or partners with means for either production or marketing. However, trying to sell your idea to a large firm could prove risky. Unfortunately, the PTO's Disclosure Document Program (DDP) offers limited protection over a two-year period. There is always the possibility of your partners' stealing and developing your idea. Once the protection period is up, they are ready to market your hard-earned creation.

The final option is to do it yourself. If you decide to go this route, you will document your invention (you can use the DDP forms enclosed) with witness signatures, but you will not send the paperwork in just yet. Instead, you'll safely store your documents and then develop and market your invention either with a partner or by yourself. The advantage here is that no one but you (and possibly your partner) has the details of your invention. You can be deep into the marketplace before anyone else has a chance to copy and compete with you. Another plus is low start-up costs. You can work out of your home and manufacture only what you sell. And having a partner can have many advantages. A partner is someone whom you can trust and talk to about the invention. Also, two heads are better than one when it comes to improving the idea, as well as manufacturing and marketing it. And you can pass responsibilities back and forth. If one of you wants to shoot a few holes of golf, the other can take care of business for the day. Finding a partner whose skills will complement yours (e.g., your electrical or mechanical and his or her marketing skills) is rare but can happen.

The main disadvantage to the do-it-yourself option is that if it ever boils down to a legal fight in court, the odds are against you if someone else has "prior art" (previous legal documentation filed with the PTO) or claims to your invention (remember our jiggle-maroo example?). There's always the possibility that a large company could still steal your idea, patent it, and compete directly with you down the road, forcing you out of business.

On the other hand, a patent is no guarantee of wealth for the inventor. Take, for example, Filo Farnsworth, who invented electronic television in the 1930s. Others had mechanical types, which produced terrible pictures, but Farnsworth used an "image dissector" (TV camera). RCA, the leading communications company at the time, was interested in Farnsworth's "TV," but the Second World War put the technology on hold. By the time television stations were in place in the late forties and early fifties, Farnsworth's patent had expired. The system was up for grabs. The only gratification he received from his efforts came two years before his death, when his image dissector allowed him and billions of others here on Earth to witness man walking on the moon. Recently, a statute of Filo Farnsworth was placed in the White House.

In this case, an invention was so revolutionary and took so much time to develop (World War II notwithstanding) that it got away from the inventor. Such an occurrence is rare, but it serves as one of the more extreme examples of why getting a patent is not always the best option.

At any rate, before you even consider applying for a patent or taking other steps to protect and develop your idea, you need to cover a few bases . . . which just happens to be the subject of the next chapter.

CHAPTER 2

Covering the Bases

It doesn't matter whether your invention is a free-energy device or a new type of doggy poopy scooper. You dreamed it up. You manifested that dream into reality. You are creative. But before you can open the door to financial freedom and possible fame, there are some important considerations that must precede your first steps along the narrow path up the mountain of success. There are three critical rules of thumb (the 3 Ds) you must follow before you even consider pursuing a patent: documentation, description, and discretion.

DOCUMENT YOUR IDEA

It could be the greatest invention or design in the recorded history of man or monkey, but if it is (at the very least) not in a two-dimensional form, it's just a euphoric, misty dream. The drawings do not have to be professional (in fact, many patent submissions with inferior artwork are granted), but they must be understandable. In addition, they must convey the system three-dimensionally. To meet this requirement you may simply draw front, side, and back/bottom views. But if you are good at mechanical drawing, a three-dimensional sketch can go a long way toward depicting the concept. Schematics and electrical block diagrams may be necessary if the system or device is technical in nature.

It's important to save all receipts pertaining to the documentation, building, and marketing of your invention and continue to record all activities related to your invention (e.g., photos, drawings, tape recordings, and conversations). Should you file a disclosure document, you may have to prove that you have not abandoned your intention to file patent within that two-year period. These records will also show that you alone have worked your idea out to a tangible conclusion.

DESCRIBE HOW IT WORKS

You must be able to describe in simple terms, the function of the system. Your putting it on paper means that it will never be lost to mankind. And remember to always copy your work! I have heard countless stories of manuscripts and patent submissions that were irretrievably lost because the original was the only existing work.

Back in the sixties, an elder friend of mine named Wayne wrote a satire on the Korean conflict, set aboard a naval vessel. It must have been hilarious because some of the stories he told made me laugh hysterically. Unfortunately, he tried to sell the novel in the early seventies. The Vietnam War was winding down, and at that time, the Women's Liberation Movement was picking up and dominating the publishing houses. Nobody was buying old war stories because *M*A*S*H* was very good and already on TV. But my friend Wayne was so discouraged that he tossed the draft. Today, it would probably sell big. Now, with his memory failing, Wayne is too old and tired to rewrite the script. Never throw anything away that you have created!

EXERCISE DISCRETION

Discretion is the better part of valor. Fact: in many large corporations, it is common practice to pirate good ideas if they are not properly protected. Protect your work. Back up all software using a ZIP drive and keep a paper printout. Always keep a copy of your invention elsewhere—at your brother's, sister's, mom's, or trusted friend's house or in a safety deposit box. And invest in a fire safe. I use a $100 Sentry safe with a combo lock. Your local locksmith can change the combination to suit you. A fire safe is a crucial piece of office equipment. Less-important files that you cannot fit into the safe can be stored in a good metal filing cabinet, which offers a limited amount of fire protection. I invested in a legal-wide, four-drawer Steelcase filing cabinet. This baby must weigh 125 pounds empty, but all that metal has a lot to protect.

Aside from theft by industrial spies (and they *are* out there), there are many other ways in which you could lose your invention. Protection must be also applied by the lips—*lack of* lip service, that is. Be discreet. There will come a time when you will have to reveal your invention to a friend or a possible investor. Be careful! We don't want to alarm you, but seriously—trust only those who have your best interests at heart! If the profit potential is great enough, many people will sell you out—sometimes even your best friend and partner. This is not intended to make you feel paranoid, but we are seldom prepared for what the universe is going to unfold before us. An old Samurai saying goes, "If all negative possibilities are reviewed and let go, then they are least likely to occur."

When a great idea is in the air, everyone wants to be in on

the action. Unfortunately, the most enthusiastic people are often the ones without capital. Some kind of "poverty mentality" overtakes normally reasonable people, and they become compelled to run with your ball. It's a bit like the "gold fever" that overtakes the character Fred C. Dobbs, played by Humphrey Bogart in the classic John Huston movie *Treasure of the Sierra Madre*. The main thing to remember when discussing your idea with others is to leave out the crucial data that will make the gismo work; never talk about the details. (Although, unfortunately, the simplest ideas can sometimes be the most valuable and are usually tough to present to investors without giving away the whole pizza. If the device is highly technical, that makes it easier not to describe.) The important thing is to leave out the one thing that your genius created that will make it happen, e.g., the exact degree of timing on a new type of engine you invented. Just as important is to have the person in whom you've chosen to confide sign a copy of the Nondisclosure Confidentiality Agreement, Form PS-9a (see Appendix I). Of course, this still won't guarantee that he will be trustworthy. Have you known this person long? Is he well known and in good standing in his community? These are important things to consider. There were times when I found myself stuck on a technical problem or a decision involving one of my inventions. Fortunately, my business partner (and coauthor of this book), Mr. Craig Herrington, is not only an old friend but is mechanically adept and a marketing expert as well. The question of trust is not an issue with us. But as a general rule, if you find yourself turning to someone for advice and opinions, be sure to have a copy of Form PS-9a ready for him or her to sign and send in your disclosure document to the PTO.

On the flip side, when you want to discuss your creation with someone you trust, it's natural to turn to family members. It is human nature to blab; we want recognition for our genius, and we all need a few ego strokes once in a while.

However, in many cases, an inventor will find that his family and close friends are nonreceptive to his new idea or invention. This is because the concept is foreign, while he as a person is familiar. Since patents are considered difficult to get, your friends and family members may think that you are wasting your time. In this case, it's important to remember that worthwhile goals always have their associated risks and obstacles; these are simply part of the inventor's path to financial freedom. So when seeking advice, be sure to listen carefully, even if the information is not what you want to hear. It may save you from making costly mistakes. On the other hand, bear in mind that no one has ever succeeded by dwelling on the negative! Take the criticism with a grain of salt and forge ahead. Nothing is more satisfying than proving others wrong with your own success.

We like what Bill Nasset, an acquaintance who is a noted inventor and marketer, has to say about the potential of making it big as an inventor:

> Also, there is the protection time of a 20-year patent to consider. Even a high-demand, flash-in-the-pan short-lived invention can be extremely profitable. How would you have liked to have had a patent on the Hula Hoop, Barbie Doll, or Slinky? Need I say more?

CHAPTER 3

Working Out the Kinks

At this point, you are ready to build the first prototype of your invention. You must have a prototype in order to demonstrate your product's feasibility before you can patent it, let alone license or market it. Building a working model of your invention may be one of the greatest accomplishments of your lifetime, so it's a good idea to work it out in your mind or on paper before you actually build it. Many times ideas that look good on paper are total flops in three dimensions. But bear in mind that not all inventions are created with a single stroke. Many times devices require a number of modifications before they are functional. Open your mind to what is already available in the outside world. You could be walking through an antique shop and find a solution or new idea just lying there on the shelf. I built a prototype inertial propulsion engine using miscellaneous parts from junk boxes I collected over the years. In my case, the machine turned out to be quite functional, but you may not get so lucky on your first try. If your model fails to work initially, don't give up!

Could you be overlooking something simple? Take a few days, or even weeks, to simply set it aside and put it out of your mind. Just forget about it! I often solve a highly technical problem by just dropping it for awhile. Sometimes I've put it aside for months—or even years—before finally coming up with the solution.

Of course when it comes to *your* invention, it's best to not take too long, as the marketplace awaits. I now use my secret weapon, dream yoga, or lucid dreaming, to solve such problems. (It would be a real shame if someone else came up with your idea and actually made it work, when all you needed to do was to sleep on it for awhile!)

Many problems are solved while one is asleep, during the dream state. As crazy as this may sound to you, it really is not. Great inventors throughout history have had that big breakthrough in the middle of the night.

Looking at your notes before falling asleep often sparks a subconscious suggestion that becomes a lucid experience while you're dreaming. I practice lucid dreaming when possible and find that many problems I cannot solve while awake are easily sorted out in a lucid dream where I am conscious that I am dreaming—or become conscious once the solution manifests itself.

DREAMING IT UP:
DEVELOPING YOUR CREATIVITY

Once we have fallen asleep it takes about two hours before we have our first dream. We usually have about five to six dreams a night. The first dream occurs because the mind must resolve the most pressing problem. First dreams are usually either violent or disturbing. This is when nightmares occur—near midnight. The second dream is less intense, usually about work or relationships. The third and fourth dreams are background events of the previous day. By the fifth or sixth dream, there is room for more leisurely events, since the pressing problems have been played out. The last dream before waking is almost always pleasurable, sometimes about flying, but often sex is the theme of the dream. This symbolizes an awakened state of mind; all senses are in high gear. This is also the best time to unwind a technical problem relating to your invention. You are well rested and somewhat lucid, but between conscious and unconscious states. If you wake up after an erotic dream (which is generally the case), go back to sleep for 20 minutes with your problem on your mind. The best dream solutions occur in the early morning hours or just before dawn or rising.

Keep a notepad and pen on your nightstand. Often I have awakened with a great idea only to fall back to sleep because I did not want to get up and search for a notepad and pen. Besides, once bright light hits your eyes, it will be another 90 minutes before you will be able to fall asleep again.[1]

Dreams can be symbolically profound. For example, in the spring of 1996, Craig Herrington and I were trying to develop an engine that would outrun its own inertia and lift off without jets, rockets, or aerodynamics. We had already made models that would travel horizontally, but they had little thrust. Using the force of rotary motion, these engines appeared to break the laws of physics, namely, Newton's Third Law of Motion (action/reaction) and the Laws of Conservation of Energy and Momentum. But inertial propulsion does not break any known laws in physics and is a working principle!

Still, you can imagine the resistance we encountered, even after demonstrating our engines before the scrutiny of the scientific public.[2] But, because there was such low thrust (not enough power to propel the engines with a load), we received little attention. So I decided we had to build a bigger, better engine—one that would lose weight and lift-off. But I did not know what to build or how.

17

I fell asleep one night with these problems weighing heavy on my mind. I awoke around 5 a.m. and remembered them, then drifted back to sleep. I glided across a vivid, dew-drenched blue-green meadow, where I came upon a 20-foot-high stone wall. It looked ancient. Moss covered some of the lower stones. Swallows took refuge in some of the upper cracks between the stones as they flitted about beneath an azure blue, cloud-puffed sky. The earth rumbled as one of the top stones slowly slid out and hovered before the wall. Its face dissolved into a brilliant golden light. Out of the end of the open stone glided out a rolled parchment scroll. It unrolled, revealing the details of an inertial propulsion system unlike anything I had ever seen. The legends were alien. I awoke, fully understanding the dangerous and beneficial impact of what I had discovered. Excitedly, I rough-sketched it out on my bedside notepad before jumping out of bed.

I spent that morning drafting out the basic design for Engine #8. Two months later, I had several patent-quality engineering drawings depicting the whole system—electrical, electronic, mechanical, and systematic: I would never again start another project without first completely laying it out in two dimensions.

Afternoon naps are very beneficial. Thomas Edison took many catnaps. On the other hand, Edison's rival, Nikola Tesla, hardly slept at all.[3] (Fortunately, in the debate between AC and DC power distribution, Telsa won out.) Although there are exceptions, most of us do need sleep and, in fact, many of us are in a constant state of sleep deficit. A nap refreshes and allows us a chance to create and discover.

Thanks to modern technology, you can purchase a "lucid dream machine." The Nova Dreamer is a self-contained pair of soft goggles with infrared sensors that detect (without direct contact) rapid eye movement (REM), which occurs during dreaming. Five minutes later, LEDs flash, indicating to you that you are in a dream. The kit includes an operation manual, an audiotape, and the book *Exploring the World of Lucid Dreaming* by Stephen La Berge, Ph.D.[4]

Imagine the problems that could be resolved in the REM or dream state. Being consciously awake while dreaming is absolutely fantastic. I have solved most of my problems just by dreaming about them.

Even if you come up with a new idea to improve the chances of your invention's working, you still may need to confide in someone for technical assistance. If you do need to bring in outside help, don't get too generous. A mere 1- to 3-percent interest in your invention could mean hundreds of thousands of dollars to a potential partner over the long haul. Whether or not you are able to *build* the device you conceived of, you are still the *inventor* and have full rights to the device. Down the road, you don't want this technical advisor to be called coinventor. If you must solicit help in building the prototype, make sure that you have already recorded and filed form PS-1b with the DDP in Washington, D.C., and have the party or partner *sign and date it in ink*. (The date is important!) Stipulate that the third party is only lending technical advice. The Nondisclosure Confidentiality Agreement provided in Appendix I, signed by your partner, is the bare minimum you'll need to protect yourself. There are better forms on the market, and many manufacturers will want you to sign one drafted by their attorneys. *Read all contracts carefully before signing them.* You don't want to sign away your rights. Don't be afraid of appearing stupid by asking questions—no matter how simple the question may seem. It's your future on the line. *But above all, read this entire book before you send anything to the PTO.*

DO YOU HAVE SOMETHING?

As an inventor, your goal is always to be ahead of the game. By the time a trend or style becomes a worldwide sensation, it is usually too late to get in on it. You want to invent something that is truly new. Beware of novelties, unless you can make and sell them fast. Many people have become millionaires selling novelties. Take the pet rock, or robot eggs (large steel ball bearings), for example—not exactly inventions based on genius, but hot flashes in terms of sales, nonetheless.

There are three big questions that you must ask yourself:

1. Has it already been invented?
2. Is it already on the market?
3. Does it exist in another form?

Getting a patent is risky business. There's always a chance that it could result in your invention's being exposed to the world before you are ready to develop and market it. True risk is impossible to determine 100 percent. Whether it's worth the risk is something you'll want to consider carefully. In any case, you will have to spend some time and money to protect your invention.

At this stage of the game, you need to do your own patent search. No, not at the PTO in Washington, D.C., but right in your own hometown. Visit stores or suppliers that would sell your product to see if they have something similar to your idea. Engage the salesperson in a few questions, such as, "Have you heard or seen a product that does this and that?" Salespeople love to talk. Don't volunteer any information on your idea (which is a compulsion for many of us). You can reap a great deal of information just by listening. The key in your search is to keep lips sealed, eyes open, and ears alert to gather as much data as you can about prior art and product feasibility.

Suppose someone tells you that he or she has heard of your idea or possible product. Find out where! It could just be that your suggestion made the person see that there is a need for your invention. The power of suggestion can sometimes cause the mind to create scenes that never happened. The product may not exist—yet!

Go to your local library and check out trade journals, magazines—any publication related to the field of your invention. (When I have a design problem, I go to the public library and scour the *Thomas Register of American Manufacturers*, which is an alphabetical listing of products made by any manufacturer worth listing in the United States and Canada. Any company listed (most of them have toll-free numbers) will be happy to send you its free catalog. Some weeks I receive up to 25 catalogs—all free, and some over an inch thick. (A good source for doing preliminary patent searches and locating anything you might need as an inventor, the *Thomas Register* (2000 edition) was available for $149 on Amazon.Com at the time this book was published.)

Don't be surprised or disappointed if your patent search turns up a number of old patents directed towards the problem your invention is intended to solve. Old ideas sometimes are great inspiration for best-selling new ideas. Not all problems are 100-percent solved by previous inventions, and even if prior art does exist, you may still patent your invention if you can make a new claim, or present a better way of solving a particular problem. For example, Alexander Graham Bell, inventor of the telephone, had his mansion air-conditioned long before the word existed. He built an asbestos-insulated duct, or air-shaft, from attic to ground floor. In the attic he had a big open box filled with ice and salt, while on the ground floor he kept the tops of the windows open just a few inches. Cold air came down the shaft to the first floor rooms. This, in turn, forced hot air out of the tops of the ground floor windows and kept Mr. Bell cool to create whatever.[5] While Bell had an answer to the air-conditioning problem, it was the wrong answer for anyone but a rich inventor with a household full of servants. The right answer didn't come along until the portable window box compressor/fan unit of three generations later.

So, if you discover that someone beat you to the punch, don't give up immediately. Your bright idea may be the right idea. Safety pins were old hat in ancient Rome, but they didn't become a big deal until American inventors perfected them more than 2,000 years later.[6]

Another, more personal, example: I was so caught up in my own work on Inertial Propulsion (a proposed safe, efficient space-drive system) that it wasn't until after I gave a demonstration to engineers and scientists at the International Tesla Society's Extraordinary Science Symposium in 1995 that I learned how many others had attempted to solve the same problem over the past 80 years—and filed more than 100 patents (in more than a dozen ways)!

Do your research! Had I been familiar with the work of many others in my field, I would have a fully functional electric satellite engine by now.

MARKETING POTENTIAL

The next important consideration is marketing potential. Inventing is satisfying, but to be a wealthy inventor, you need to invent what sells!

Make a realistic appraisal of the actual need or demand on the part of the public. It's best to try to develop something really useful—a product that will have a long-term effect on the consumer because it fills a basic need rather than an emotionally based desire. Once you find that invention, and develop it properly, you will find yourself very wealthy.

On the other hand, necessity is no longer the mother of invention; dissatisfaction is. It's what breeds that unquenchable desire to fill our space with something new and unique that not only makes a statement about us but also satisfies us in some way—or allows us to be "innovative" by being the first, or only, one of our friends to have such an interesting item. This psychological motivator is known as "fear of loss" in sales jargon. On a personal note, I believe that in general, we, as Americans, have overwhelmed ourselves with so many material doodads that we really do not know what we want. So we try hard to find satisfaction by buying more stuff. This is why the market can be so unpredictable.

One must not forget the fad-toys of the past that have filled emotional needs and become a part of American culture such as the yo-yo (now with LED lights and backlash springs) and even the Hula-Hoop (which made a short reemergence in the early nineties and is still seen in stores today, although the original inventor is no longer receiving royalties, of course). Still protected by design patents and trademark rights to this day are Slinkys, which have gone plastic, and Barbie dolls, which continue to evolve with fashion and are destined to be forever Americana.

Ask sales clerks if they would want something on their shelf that does "this and that." Could they move it? Would it sell? Talk to people in the trade. Can they see a use for your invention? (As always, discretion is important; don't reveal anything!)

Once you've completed this informal market research, you must take a critical approach to your invention. Consider the following:

- Uniqueness: Does it appeal to a large portion of the public or have a place on the market? Does it fill a need and are the benefits obvious?
- Limitations: Is it a seasonal item or one with a short shelf-life? Is it safe?
- Durability: Is it a flash in the pan or would it prove useful in the long term? Is there potential for repeat sales?
- User life: Is it durable goods or consumable goods?
- Manufacturing requirements: Will it require special molds, machining, or equipment?
- Manufacturing location: Can it be made locally, or must it be made out of state or country?

- Design simplicity: Is it simple and cheap to make, or is it high-tech?
- Usability: Can the consumer readily learn to use it with minimal training?
- Packaging requirements: How much packaging will your product require? Is it easy to package?
- Consumer cost: What is the retail price? Can it be reasonably priced?

Make price comparisons between your invention and existing products that relate to it. It's important not to overprice the unit, or you may have a difficult time marketing it to a manufacturer. Market feasibility carries a lot of weight in the inventor's handbag. For example, you may have designed a powerful ion space drive that uses cesium metal and water, but if the price of cesium is higher than the price of rocket fuel, then you are wasting your time—unless you find another way to power the drive, to make it work within our atmosphere. The point is, people will buy it only if the price is reasonable.

The common rule of thumb is the 1:10 ratio: calculate how much it would cost in parts to build 100 units. This is because most suppliers give discounts in 100+ parts lots. Add up the costs of all the parts and divide by 100. This will give you the average discount cost of one unit. Now, to manufacture that unit, you must pay a labor force to build, package, and ship the product. Of course the distributor will want his piece of the action. Simply put, multiply every dollar you spend to build a unit by 10, and that will be the consumer's price. (As a general rule, the 1:10 ratio works for small item production runs, not prototypes. Prototypes typically cost more.)

You may need to consult a marketing company for a professional evaluation. There are hundreds of thousands (or even millions) of dollars to be made with a good product that is properly advertised.

However, beware of marketing companies that want money up front. Stay away from "inventors' clinics" and "invention brokers." As a teenager in the late 1960s, I sent a detailed description of a magnetic padlock to an "invention broker" and never heard from them again, except to see a magnetic padlock on the market three years later, advertised in my paid *Popular Mechanics* subscription. It has been on the market ever since. My hot-air comb of the early 1970s eventually showed up on the market as well, the result of my having trusted another agency. So I know from painful experience that secrecy is tantamount.

FOR WHAT IT'S WORTH

There comes a time when you must face the facts. You must be honest with yourself. Is the idea unique? Is there anything like it on the market? If your idea has already been patented, once was, or already is on the market, don't waste hundreds of dollars on the patent process. Find another idea that will sell and work on it. At the very least, you are also receiving a firsthand education; the skills you picked up doing your informal patent search will remain with you.

On the other hand, there may even be enough difference between your invention and a similar patented product to justify your applying for a design patent. And keep in mind that even if someone else has produced and sold your exact idea, that doesn't mean that you cannot market it! If you can improve on the device, and if the patent has expired, the invention is up for grabs—you have the right to market it.

Your informal marketing search should give you a good idea whether your idea is unique enough to be worth your time. If you decide that it is, then, you must be prepared to commit to the process once you have started. Getting wealthy on an invention takes effort. You'll have to spend time on the process, and it will involve a great deal of stress. The situation is a bit more complicated than it was in Thomas Edison's time. In fact, there were many times when I nearly gave up, but I got back in the saddle. The possible rewards are just too great to allow yourself to become so discouraged that you give up.

NOTES

1. www.lucidity.com (Palo Alto, CA: The Lucidity Institute, Inc., September 2000).
2. *Inertial Propulsion: The Pendulum Test.* (Galesburg, IL: Appletree Press, July 1995). An extraordinary science video available for $55 (plus $5 shipping) through the Appletree Press Web site (www.misslink.net/wedgees).
3. Thomas C. Martin, *Dr. Nikola Tesla . . . His Inventions, Researches, and Writings* (Colorado Springs, CO: ITS, 1987).
4. Available through The Lucidity Institute, Inc., 2555 Park Blvd., Ste. 2, Palo Alto, CA 94306.
5. Alan Montague, "What the Patent Law Means to the Inventor," *Science and Mechanics* (September 1962): 28.
6. Ibid.

CHAPTER 4

Free Money!

As an inventor, you will face many challenges that can be sources of discouragement along the road to success. Lack of inspiration and mechanical glitches that prevent your idea from working as you intended it to are just a few. Lack of funding is another, and it can be a biggy. However, if you can find an organization to back your idea financially, you can become encouraged very quickly!

There are many organizations that are willing to provide financial backing for ideas they believe to be worthy causes. They are very selective and give out only a certain amount, while managing their funds carefully so that they can continue to give out more each year. Nobel Prizes are an example of how an organization can become larger and more generous each year. The money Nobel has is invested wisely and grows, which allows the organization to give out more prizes and greater amounts as time goes on. The total amount awarded is never more than would safely allow the Nobel fund to grow larger through the interest earned on investments.

WHAT ARE GRANTS?

First, let me dispel some myths about grants. If you see an ad offering to get or help you get a grant from some source, there is a 99- percent chance it's there for the sole purpose of taking your money. There are a few legitimate companies out there that can assist you in obtaining grants, but they are few and far between. And make no mistake, all organizations that give grants exist to promote their cause. You will see shortly what I mean by this.

Grants come from many sources, but these sources fall into two main categories: private and government. Every arm of the government has its own grant program, some of which are covered below. I will concentrate on the private sector because it is the most favorable to the individual.

Private foundations hand out billions of dollars per year to companies and individuals. Very similar to licensing, securing grants is a field of its own, and there are dozens of books available that will guide you through this complex money machine. I will give you some guidelines on how to proceed with your research on grants, but you must realize that it will take time and study.

Grants fall into three major categories: educational, scientific, and assistance. In most cases, applying for a grant requires a written proposal. Some proposals are simple, but most are long and complex and must be followed to the letter. I recommend that you visit your local library or bookstore and get some books on writing proposals for grants. Such books will go a long way toward guiding you through the maze. Also at your local library, you will find *The Foundation Directory*, which lists all the grant sources in the United States and some other countries. This valuable reference lists the name and type of each foundation, where it is located, who to contact, donors, financial data, purpose and activities, fields of interest, limitations, application information, staff, and other miscellaneous information. This is the same source used by all those people and companies trying to sell you information on obtaining grants— information that you can get yourself for free!

Grants are a great place to find funding for projects, but I do not want to get your hopes up. Grant money is set aside for very specific and specialized purposes and is not handed out as freely as many try to make you believe. On the other hand, it is definitely worth looking into and pursuing if your project meets a particular foundation's criteria and you are prepared for the mountain of paperwork you will need to deal with.

As one example, the United States Department of Defense (DOD) invites citizens to apply for grant moneys for inventions that address its particular needs. This is a great resource for ideas as well. What a concept! You could invent something that the government needs and will pay for developing, which you know will sell and make you lots of money!

APPLYING FOR GRANTS

You do not even need a "legal" patent to be granted funds for research and development on your invention, provided that you have a functional prototype that addresses the needs of the institution offering the grant. If your project is awarded funds for development, then you can afford to seek out a patent attorney and file for a patent on your invention. Besides, it is not wise to file before you have worked out the bugs. The funding agency will grant you funds, provided that the project is feasible, shows promise, and meets the specified cost parameters.

Some of these parameters are quite generous: Phase I of a DOD grant allows for up to $100,000 for ". . . a period not to exceed six months (nine months for the Air Force)." Phase II awards shoot up to from $500,000 to $750,000 for a period of no more than 24 months. At the phase II level, the prototype must be well developed for presentation.

At the Phase III level, you are expected to find funding in the private sector. But wow, what a break! You have a fully developed, paid-for product—and the government wants it. Investors will beat down your door to talk to you.

JUMP STARTS

Listed first are the Department of Defense's Small Business Innovation Research (SBIR) and the Small Business Technology Transfer (STTR) programs, because they offer grants that are generally easier to get than those available from some of the more elite organizations listed.

DOD SBIR/STTR Support Services
ATTN: SBIR/STTR Mailing List
2850 Metro Drive, Suite 600
Minneapolis, MN 55425-1566
E-mail: SBIRHELP@us.teltech.com

Ask to be put on their "DOD SBIR/STTR Mailing List." Include your daytime telephone number(s) along with your address. Every six months, they will send you a catalog of their topic descriptions. Find the topic that your invention best relates to, and carefully follow the application instructions. If you get turned down the first few times, don't fret. Keep trying. Each six-month period provides new topics from which you can chose, and who knows, you just might have what DOD needs. Mail in the DOD Mailing List request provided in Appendix IX.

If your invention pertains more to the civilian crowd and promises to have a significant impact on society, then you might try the National Science Foundation (NSF). These people are picky, but they are worth a shot. They usually like to see "M.A." or "Ph.D." after your name.

National Science Foundation
4201 Wilson Blvd.
Arlington, VA 22230

Ask for a copy of their "Guide to Programs" and "Grant Proposal Guide."

The Office of Technology Innovation is a function of the U.S. Department of Commerce. Ask for a copy of their "Invention Evaluation Request" form. They are especially interested in inventions that efficiently use, transform, or create energy.

Office of Technology Innovation
National Institute of Standards and Technology
Gaithersburg, MD 20899-0001
E-mail: innovate@enh.nist.gov

Be particularly careful how you describe your invention. Get as close as you can to using "patent language" (see Chapter 5). This will indicate that you did your homework and probably have applied for a patent. You may use just the major claims outlined in your documentation. (Unfortunately, videotapes are not accepted by this or any of the other foundations listed.)

CHAPTER 5

Get a Patent

Suppose you are not successful in getting a grant. Chances are you will not have the financial resources that are often needed to get a product on the market. Another option is to seek investors to give you the financial backing you need. But in seeking investors, you take a risk. You must reveal your idea to them before they will agree to fund your product's launch into the big time. What's to prevent them from stealing your idea and marketing your product themselves? For long-term legal protection (beyond the two years of limited protection the DDP provides), the most obvious solution is to file for a patent. Traditionally, this is the road most traveled. It is a fairly lengthy process, and it can also get expensive—especially considering it is not an iron-clad guarantee of anything. But it is one option, and to make informed choices, you must understand all of your options clearly. So let's look at what is involved in getting a patent.

IS THE IDEA PATENTABLE?

Before you rush to prepare and file your patent application, pause for a moment and think—have you really worked out your invention?

Some time ago, an amateur golfer decided to investigate the novelty of what he believed to be a revolutionary golf ball, one that could not be lost easily and, thus, would nearly eliminate one of the game's most aggravating aspects—time spent tracking down missing golf balls.

The would-be inventor proposed to coat the golf balls with a compound that would emit smoke for a few minutes after impact. The golfer could find his ball simply by following the smoke trail! After several hours of poring over copies of prior patents on golf balls, he was assured that none existed for a ball built to emit a smoke trace.

He decided to apply for a patent and consulted a patent attorney for assistance in preparing his application. The attorney—knowing that in order to accomplish its intended purpose a patent application must contain a detailed explanation of how the invention is constructed—asked his new client what compound he had found satisfactory for the purpose of giving off smoke when a golf club hits a ball.

"Oh," came the reply, "I haven't found a compound yet that will do it!"

"In that case," the attorney said, "you haven't any invention at all. You have nothing but an idea for something that would be a good thing if and when someone invents it."

You can't get a patent on a good idea. Patents are granted only for the novel structure designed to carry out an idea.[1]

The above example typifies one of the most common misconceptions about patents. Often, I hear the phrase, "He patented the idea." or "I have a great idea for a patent!" Patents are issued for an apparatus, machine, manufacturing process, compound, chemical, and so on—i.e., the vehicle

through which an idea is expressed or made functional. As
Aubrey McFadyen explains in his article "Have You Fully
Worked Out Your Invention?," "Rather than the idea, it is the
physical structure that is patentable, or a tangible procedure
which incorporates and gives effect to the idea."[2] According to
the patent system, an "invention" is the solution to a problem
rather than the mere recognition that it exists. States
McFadyen, "Patents are designed to reward those who have
made a tangible contribution to mankind; patents cannot be
granted for wishful thinking."[3]

WHAT ARE THE CRITERIA?

Anyone may file for a patent—including deceased or
mentally incompetent persons, as long as they can apply
through a proper representative! (In cases where an inventor
dies, becomes mentally ill, or cannot be located, the patent
law has a provision under which others may file on his
behalf.) In a nutshell, there are only four criteria your invention must meet. If it passes all four of the following "tests,"
then the PTO must issue you a patent:

1. Is it unobvious? (Does it create previously unknown
   results?)
2. Is it novel? (Does it use a combination of previous features or make new use of an old feature, or is it completely
   new?)
3. Is it useful? (If you were to market the product or system,
   would people use it?)
4. Is it in a statutory class? (Does it fall into a category
   wherein it can be classified as an existing, viable technology? Abstract ideas cannot be patented. But if you take an
   abstract idea and make it work as a physical device, then
   it is patentable.)

Memorize these four criteria before doing your patent-ability search (more on this later).

If your invention does not pass these four tests, don't waste your time and money proceeding further. Come up with a new invention!

HOW MUCH DOES IT COST?

Patents are expensive. (In Appendix V are the latest price listings for the various services available from the PTO.) The average base cost is around $4,000. Also note that every 3.5, 7.5, and 11.5 years you must pay a maintenance fee.

If you hire an attorney to do a patent search, add on his fees. Some patent attorneys charge between $100 and $150 per hour. And unfortunately, there is no guarantee that you will receive a patent.

In many ways, this may be fair: patent attorneys do not know before making a patent search whether or not your idea has been patented previously. They need to feed their families too, and searching through hundreds of files is time consuming. But interestingly enough, the PTO recently computerized its filing system, making it simpler to find various patents. However, specific claims contained within individual patents have not been entered into the system, since this would be a Herculean task, even for such a large organization as the PTO. The last I heard there were more than 5.1 million patents on file.

Some inventors vie for doing a patent search themselves via the Internet. It's not possible to do an as-of-current-date search, but everything else you need to know about patents and patent law is on the 'Net. You can even contact UPS directly on the 'Net to ship you hard copies of all patents related to your invention, provided you know the patent numbers. NOTE: If you wish to review a patent and know its seven-digit number, you can order it directly from the PTO for just a few dollars. The PTO takes only credit (or debit) card orders. Call (703) 305-8716 to place an order.

Appendix VI is a Patent and Trademark Depository Library Program (PTDLP) list. Included are all the major libraries in each state (with their telephone numbers) that have patent searching capabilities. With a little practice and a lot of patience, you can do your own patent search—saving yourself thousands of dollars in attorney's fees. And keep in mind that should you find an existing patent that is similar to your idea, this doesn't preclude you from patenting your idea. Remember criterion number two: if your invention is new or novel, it can be patented.

UTILITY OR DESIGN PATENT: WHICH DO YOU NEED?

The PTO grants three basic kinds of patents: utility (formerly known as mechanical), design, and plant. The application process is the same, but unless you are someone who invents, discovers, or asexually reproduces a distinct new variety of plant, you would not be interested in the third type. So which do you need—a utility or a design patent? Let's take a look at the differences between the two and how much protection each affords you.

What's the Difference?

Because utility patents are by far the most important kind—both in terms of the number granted and their commercial value—the term "patent" is commonly used alone to refer to what is technically a utility patent. Utility patents cover processes, electrical and mechanical apparatuses, and chemical compositions, and their function is to protect the basic physical structure and principles of such inventions. The grant, which runs for 20 years, includes drawings showing the structure of the invention, together with an elaborate specification fully explaining the invention's scientific principles, and claims defining precisely what features are covered by the patent. Outright ornamentation plays no part whatever in mechanical patents. Appearance is important only when it serves a functional purpose, as in a drill bit, where the shape of the bit governs its cutting efficiency, or in an airfoil, where the curvature or shape affects its lifting or aerodynamic properties.

Design patents, on the other hand, are based on the appearance of the item, rather than the basic physical structure and principles of operation. By comparison, they make up approximately a tenth of the total weekly issue of patents. If the invention is suited to it, a design patent offers several marked advantages over a utility patent: the application is simple compared to filing for a mechanical patent; the patent can be obtained with relative promptness; and the government and attorney fees for a design patent run only a fraction of the cost of a utility patent.

Under the law, "any person who has invented any new, original, and ornamental design for an article of manufacture" may obtain a design patent on it, subject to the same general conditions that govern the issuance of utility patents. The shape of a new combat knife, for example, might warrant a design patent. Let's examine the three key aspects of this law.

Ornamental
The beauty and ornamentation required for design patents is not confined to the so-called esthetic or fine arts of painting and sculpture, but in general extends to all types of manufactured articles. With regard to manufactured items, the word "ornamental" generally refers to something that presents a pleasing appearance, although, because individual tastes vary, some people may think that an article looks grotesque, bizarre, or even ludicrous. Nevertheless, the item can still be considered ornamental within the meaning of the statute. Impressions imparted by the design may be complex

or simple; thus, one design may leave in the observer a mixed impression of gracefulness and strength, while another design may leave the impression of strength alone. But whatever the impression, it must convey to the observer a sense of uniqueness and character to be a patentable design. The majority of design patents cover the general lines and contour of the article, such as in automobiles and vacuum cleaners. Surface ornamentation, as in pictures and carvings, occurs primarily in connection with such things as jewelry, dishes, silverware, wallpaper, and coffins. These types of items represent a minor sampling of current design patents.[4]

Invented

As with utility patents, design patents must be based on inventive ingenuity. The so-called invention in designs is more similar to what might be termed the originality in a painting or statue than it is to the invention in a method or machine. Again, since there is a wide divergence of opinion, even among artists, with regard to what constitutes ornate or attractive, the courts and the PTO take a practical and liberal view as to a design's patentability, taking into consideration the design's appeal to the eye of the general class of persons for whom the design is intended. The working rules laid down by the courts in design cases hold that the inventor must do more than make routine changes, such as the substitution of curved for straight portions or the mere selection of vivid colors for his design. Thus, a design patent for a candle having a square cross section was declared invalid, the court remarking that the public has the right to make any article round or square or in any other "standard form."

Nor will omitting a few minor details of a prior ornamental design entitle you to a patent on such modification. However, the reassembling and regrouping of familiar forms and decorations may constitute a patentable design. Thus, where various design features of a hunting knife appeared in several, scattered prior patents—one disclosing the blade, another the handle, and so on—the courts held that the combination of these various features resulted in a unitary and ornamental design exhibiting the quality of invention.

Article of Manufacture

While design patents tend largely toward styles of apparel and jewelry, household articles and appliances, patterns for fabrics and rugs, and furniture and furnishings, they can also include hardware and carpentry and machine and garden tools. In fact, there are more than 100 different broad titles or classes of goods that have been issued design patents, including virtually every article found in commerce. Articles covered by design patents range from simple things like pins, bottles, and packages to automobiles, airplanes, and locomotives.

Although the statute states that "design patents may be issued for any article of manufacture," the courts have imposed several limitations on what types of article may be patented. For instance, the courts have held that a cogwheel, normally covered with grease and hidden in a gear case, is not subject for a design patent; for, by reason of its obscure use, its appearance in no way delights the purchaser. The same holds true for a horseshoe and a drill bit for wells. But the courts have ruled that a road-paving machine has eye appeal and therefore is a proper subject matter for a design patent.[5]

A design patent cannot be resorted to as a subterfuge to protect purely mechanical features. Mechanical structure can only be protected by a utility patent. Thus, a design patent on an automobile bumper having special mounting brackets, even though the brackets had a distinctive appearance, was held invalid, since the things that gave the bumper sticker distinctiveness were purely utilitarian. However, a design patent on an attractive license plate holder was sustained. Likewise, design patents for automobile tire treads have been denied on the basis of the mechanical function of the tread as well as their inherent lack of aesthetic value (by analogy to the horse-shoe case), whereas design patents for sidewall ornamentation and the overall appearance of the tire have been sustained.[6]

Ordinarily, applications for design patents simply state the title and nature of the article and contain no description. But where a description would give a better understanding of the design or the article to which the design is to be applied, such a description is permissible and should pertain to, or outline, the illustrated design.

Another key difference between utility and design patents pertains to claims. Whereas the claims in a utility patent must define in words the structure and scope of the invention and usually comprise several recitations of the invention, only one claim is permissible in a design patent. Since it would be impossible to paint a verbal picture of a complex design, the claim in a design patent takes the invariable form, "I claim the ornamental design for a (name of article) as shown." Since only the design "as shown" is covered by the patent, it is important for the application drawing to show clearly and precisely every detail of the design.[7]

Patent or Copyright?

While we're on the subject of design patent drawings, let's clarify an issue that is often the subject of some confusion: a copyright is not the same as a patent.

To begin with, you cannot get a copyright on an invention. A patent covers substance, whereas a copyright covers form.

You can get a copyright registration on a set of blueprints showing how to construct an invention, but this copyright protects only the blueprints, and not the invention itself.

That is, the copyright covers the form in which you make the drawings; it does not cover any use that may be made of the structure revealed in the drawings. So, you can sue an infringer for reprinting your blueprints, but you cannot sue him for making the invention itself. In fact, if you have a mechanical invention, not only would a copyright on the drawings fail to protect the invention, it may even bar you from securing a valid patent unless you apply for your patent within one year from the publication of the copyrighted drawings. After an invention is documented, the PTO allows the inventor a one-year grace period to apply for a patent. After this period, the documented invention becomes "prior art," and the PTO will not issue a patent on published material, even though it is not written down on a patent application. Your invention then becomes *public domain*, i.e., it is published. But this law also works in reverse: you can publish copies of patents and sell them by subject matter in book form, but you can't copyright the material. It is a law designed to prevent monopolies on paper: you can't tie up progress, which is why the PTO was established in the first place—for progress.

Copyrighting Drawings

One concern inventors often raise involves the copyright of drawings. Let's start with this simple fact: what you draw is your own. By common law, you actually have it in copyright; if you have the original, legally it is yours. This means that I can't peek over your shoulder and copy it. It also means that if you bind me to secrecy and then show it to me, I can't copy it. And furthermore, if I agree to pay you to use the idea set forth in the drawings, then I can't use the idea without paying.

Now, since a copyright gives you nothing but a ticket to sue for infringement on your drawings, and since what you (as an inventor) seek is the right to sue for infringement on your idea, a copyright is of little practical value.

For instance, if I had a basic patent on the wheel, you would be infringing by making a cogwheel. But if I had a copyright on a print of the wheel and you printed a drawing of a cogwheel, you would not be infringing.[8]

In fact, with or without a copyright and with or without a patent, there are many instances in which, if a manufacturer agrees to pay you for using your idea, you have a contract with him and can sue him if he violates it. However, each case varies, so consult a lawyer on this.[9]

If, after all of this, you still think you need to secure a copyright, there is a two-page copy of the FORM TX copyright application in Appendix VII. (If you wish to copyright printed material, copy both pages onto one page since the form is actually on one sheet of paper.) NOTE: You may also order free of charge, a few copies of the Form TX directly from the Copyright Office. The phone number is (202) 707-3000. At the time this book was published, the price for a copyright was $30.

THE PARTS OF A PATENT

When filing for a patent, you are pledging an oath that you believe you are the first and true inventor of what you claim to be your invention, and you must sign and swear to that oath.

A patent is a contract between the inventor and the people or government, and, like all contracts, it is based on a consideration or value passing from one party to the other.

The inventor turns over a complete explanation of his invention so that the public can use it freely—on expiration of the patent. On the other hand, the people (government) pay the inventor by guaranteeing him exclusive use of his invention for the duration of the patent (now 20 years).

A patent is much like an ordinary deed to land. The jacket, or folder, signed by the proper authorities has a standard form printed on it. The form states that, upon the PTO's examination of the patent, the inventor has been "adjudged justly entitled to patent" on the invention disclosed in the specification "hereunto annexed" and grants to him the exclusive right to make, use, and sell the invention for 20 years.[10]

5,050,810
ONE-PIECE PULVERIZING ROLLER
Robert L. Parham, 1675 Larimer St., Ste. 625, Denver, Colo.
80202
Division of Ser. No. 464,870, Jan. 16, 1990. This application Jun.
18, 1990, Ser. No. 539,574
Int. Cl.^{5} B02C 15/14

U.S. Cl. 241—293   4 Claims

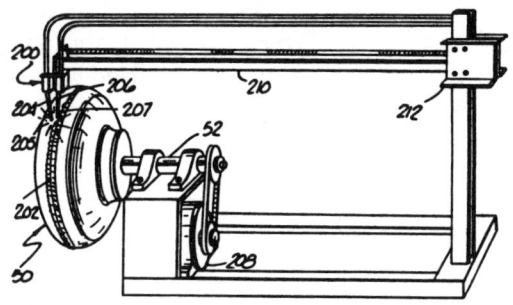

1. A pulverizing mill roller for mounting on a pulverizing
mill shaft assembly, comprising:
   (a) a roller body having a circumferential portion, said cir-
     cumferential portion having a hardness of less than 300
     Brinnell Hardness; and
   (b) an overlay of welding beads welded to said circumferen-
     tial portion.

The Specification

The "hereunto annexed" specification referred to on the
jacket of the patent is a detailed description of the invention. It must describe completely the machine, manufacture, manner of construction and operation, composition of
matter, improvement, or, in the case of a process, the novel
procedure to be followed. Whenever applicable, it must also
explain the operation and any principle involved. In the
case of an improvement, the specification must particularly
point out the parts related to the improvement.[11]

The specification should be arranged in the following order:

1. title of the invention
2. brief summary of the invention
3. brief description of views in drawing
4. detailed description
5. claim or claims

Each specification closes with one or more claims, each of which highlights various features of novelty, thus setting out the scope of the invention.[12] The claims thus correspond with the boundaries located in land deeds. (And much like a house, a patent can be bought outright or licensed and leased in favor of rent or royalty payments.)

Claims

For the inventor, claims are the most vital part of the entire patent application, since they define in critical wording the scope of his invention. Yet the claim is probably the most misunderstood aspect of the patent. Surprisingly, even experienced inventors are often ignorant about claims.

Our common notion of a "claim," such as an advertising claim or a particular promise, does not apply here. A patent claim is an exacting description. Like a mining claim, it describes that particular piece of territory that you allege you are the first to discover and call your own. (As a matter of fact, when you claim the south 40 of such-and-such a section, you get what is called a land patent, so the comparison is fairly close.)[13]

The Breadth and Language of Claims

When writing up your patent claim, you want to avoid making advertising-type claims. In other words, don't "claim" that your invention is the "greatest thing since Vegamatic." Instead, simply make a clear, factual statement of what you think you

5,051,735
HEADS-UP DISPLAY SYSTEM FOR A ROAD VEHICLE
Yoshimi Furukawa, Saitama, Japan, assignor to Honda Giken
Kogyo Kabushiki Kaisha, Tokyo, Japan
Filed Sep. 23, 1988, Ser. No. 248,881
Claims priority, application Japan, Sep. 25, 1987, 62-240110
Int. Cl.⁵ G09G *3/02*

U.S. Cl. 340—705   **11 Claims**

1. A heads-up display system for a road vehicle, comprising:
a sensor for detecting motion-related information on the
  vehicle while the vehicle is being driven along a road;
computing means for predicting the future course of motion
  of the vehicle according to the motion-related information
  detected by said sensor and parameters associated with the
  vehicle;
display means for displaying the information within the
  normal road view of a driver of the vehicle, the information including a limit turning trajectory which gives rise
  to a certain limit lateral acceleration when the vehicle
  moved along the limit turning.trajectory at a current
  speed of the vehicle; and
conversion means for converting the predicted motion of the
  vehicle into an image display on said display means so as
  to permit the driver to see said image display in a certain

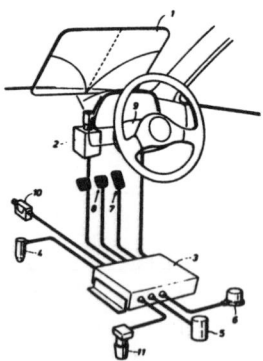

relationship with his view of the actual road ahead of said
vehicle.

have invented (e.g., "In the device of the class described, a threaded linkage rod secured by a locking washer and nut, and extendible means for producing a compulsory motion of the frick-a-ma-frack.")

You also want to avoid "picture claims"—which specify every screw and bolt in your invention. Since every word in a claim means exactly what it says, and every element it cites must be present to establish an infringement, all a would-be infringer must do to avoid being subject to legal action for patent infringement is omit one of the minor parts of the picture claim. The point is, the more all-inclusive you want your claim to be, the less detail you should cite. In a sense, you want to boil down what you have already said in your specification to the fewest possible words. And not only should they be few, but they should be broad in their meaning.

Suppose you were the first person to build a wooden box and you applied for a patent on it, claiming side, top, and bottom walls and nails passing through the edge portions to secure the parts together. Under these claims, if someone else made a box but used screws or dovetail joints to hold the parts together, he would not be infringing on your patent. If you want your patent to cover screw or dovetail fastening methods, you must draw your claim in broader language, such as, "A box consisting of side, bottom, and top walls and means securing the edge portions of the adjoining walls together.[14]

You cannot be too vague about what you are claiming. For instance, your specification may have mentioned a "spring," but in your claim you should refer to this as "expandable means," so that it will also cover a rubber band. This would protect you against someone trying to get around your patent by substituting a rubber band for the spring.

There is a fine line to walk here, however. Because while you want to be as vague as possible, you must avoid being too broad in your description and thereby monopolizing an idea. Suppose, for example, that you are the inventor of the house.

5,050,240
AIR CUSHION HELMET SUPPORT AND VENTILATION
SYSTEM WITH AIR PRESSURE REGULATOR
James L. Sayre, San Jose, Calif., assignor to Kaiser Aerospace
and Electronics Corporation, Oakland, Calif.
Filed May 14, 1990, Ser. No. 523,449
Int. Cl.⁵ A42B *3/12*

U.S. Cl. 2—6   **24 Claims**

1. A bladderless fluid cushion helmet comprising:
attaching means for positioning and holding said helmet
    relative to a wearer's head;
a sealing means, mounted proximate to and extending along
    a periphery of said helmet, for defining a volume between
    a wearer's head and an inner surface of said helmet;
an inlet port, coupled to a fluid supply, wherein a compress-
    ible fluid is supplied through said inlet port to said volume,
    and wherein said helmet is supported by the compressible
    fluid contained in said volume; and
a fluid pressure regulator disposed between said inlet port
    and said fluid supply to control a pressure of the fluid
    contained in said volume.

You might enter a claim such as, "A structure designed to provide protection from the elements." This, however, would be too broad, since it would cover the Indian tepee, the Quonset hut, Buckingham Palace, and the pup tent. Such a claim would allow you to collect royalties on everything from patio umbrellas to the Taj Mahal, and thus, it would not be allowed.

Instead, if what you have actually invented is the house, your claim should read something like this: "A structure affixed to the earth, consisting of a foundation; outside walls set upon the foundation to form the shell of the house, said outside walls

5,051,494
THERMOSETTING RESIN COMPOSITION
COMPRISING AN AROMATIC AMINE
Keizaburo Yamaguchi, Kawasaki; Yoshimitsu Tanabe; Tatshuiro
Urakami, both of Yokohama; Akihiro Yamaguchi, Kamakura;
Norimasa Yamaya, and Masahiro Ohta, both of Yokohama,
all of Japan, assignors to Mitsui Toatsu Chemicals, Inc.,
Tokyo, Japan
Division of Ser. No. 254,701, Oct. 7, 1988, Pat. No. 4,937,318.
This application Apr. 19, 1990, Ser. No. 511,235
Claims priority, application Japan, Oct. 8, 1987, 62-252517;
Nov. 10, 1987, 62-282048; Dec. 9, 1987, 62-309426
Int. Cl.^{5} C08G *73/10;* C08L *33/24*
U.S. Cl. 528—422   7 Claims

1. A thermosetting resin composition comprising about 100
parts by weight of N,N′-4,4′-diphenylmethanebismaleimide
represented by formula (e):

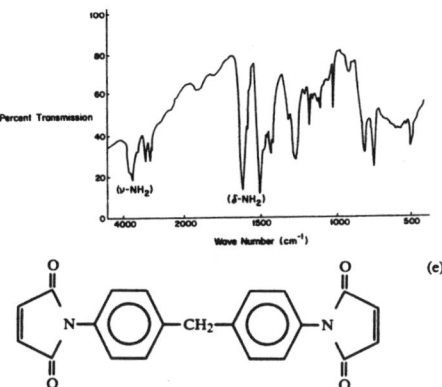

(e)

and from about 5 to about 100 parts by weight of an aromatic
amine resin comprising a mixture of aromatic amine com-
pounds represented by the formula (a):

wherein A represented a phenylene, alkyl-substituted phenylene, diphenylene, diphenyl ether or napthylenyl group, R^1
represents a halogen atom, a hydroxyl group, a C_1–C_4 alkoxy
or a C_1–C_5 lkyl group, l is 1 or 2, m is 0, 1, 2 or 3, n is an integer
from 0 to 300 and when m is 2 or 3 said R^1 groups are thes
ame or different and two R^1 groups may join to form a 5- or
6-membered alicyclic moiety.

5,050,480
TRIGGER ASSEMBLY FOR A FIREARM
C. Reed Knight, Jr., Vero Beach, and Eugene M. Stoner, Palm
City, both of Fla., assignors to Kniarmco Inc., Vero Beach,
Fla.
Filed Dec. 8, 1989, Ser. No. 447,601
Int. Cl.[5] F41A *19/35*
U.S. Cl. 89—147   **12 Claims**

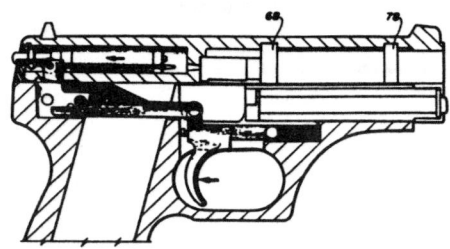

1. A semi-automatic, double action only firearm comprising:
an elongated frame having an axial slideway therein and a
    grip portion,
an elongated slide mounted on said frame for movement
    along said slideway between a slide battery position and a
    slide recoil position,
a firing pin assembly including a firing pin carried in said
    slide for axial movement between a retracted position and
    a firing position through a safety position, and
a trigger assembly including:
a trigger that reciprocates axially in said frame,
a sear reciprocated axially in said frame by said trigger, and
means that permits said sear to engage said firing pin when
    said sear moves away from said firing pin firing position,
    but prevents said sear from engaging said firing pin when
    said sear moves toward said firing pin firing position, said
    sear comprises an elongated flat strip member having a
    first leg extending from its distal end plus an upper second
    leg and lower third leg extending from its proximal end, a
    firing pin engagement lug that extends laterally from said
    second leg, and a cam surface on said third leg and, said
    frame carries a roller cam that engages said cam surface to
    cause said sear to pivot downward as said sear approaches
    the end of said movement of said sear away from said
    firing pin firing position.

being pierced with apertures for the admission of air and light; a floor, with partitions set upon said floor to divide the enclosed area into rooms; and a roof mounted overall."[15]

Your claim is essentially "A structure . . . consisting . . . of outside walls . . . and a roof . . ." Such a claim as this tells what you have invented—nothing more, nothing less. Note that you do not say what kind of structure it is; it might be brick, wood, stone, steel, adobe, or glass.[16] Neither do you claim things that you have not invented. You do not claim a fireplace or furnace, hot air ducts, plumbing, air conditioning, window panes, electric lights, gas ranges, or the kitchen sink. These are all extraneous to your invention. Including such elements in the claim would make it mean less. Again, if you say too much, you limit the scope of your claim.

Not only will the PTO refuse to issue a patent when the applicant seeks to monopolize an idea, but also where the effect of the patent would be to cover a principle or law of nature. And the courts agree. When Morse obtained his telegraph patent, one of the claims read, "I claim the force of electromagnetism, however applied, in the transmission of intelligence." This claimed purported to cover only a principle or idea: the force of electromagnetism as applied to transmit intelligence. It did not say how this force was to be applied. Morse gave no details of structure or law of operation within the claim. The U.S. Supreme Court ruled that this claim was invalid.[17]

Bear in mind that although discoveries of principles and facts of nature cannot be patented as such, the unobvious application of such a principle or fact may be patentable. For example, the patent on the use of ether as an anesthetic to produce unconsciousness during surgical operations was invalidated by the court on the ground that it was nothing more than the discovery of a fact of nature. Ether's anesthetic value had long been known by the time the inventor applied for this patent, and his application contemplated no unobvious application. However, since then, many patents have been granted

---

5,050,337
TRAP SETTING DEVICE
Nathan J. Moore, and Arthur R. Moore, both of 2263 County
Rd. H, Deer Park, Wis. 54007
Filed Feb. 12, 1990, Ser. No. 478,612
Int. Cl.⁵ A01M *23/28*

U.S. Cl. 43—97   18 Claims

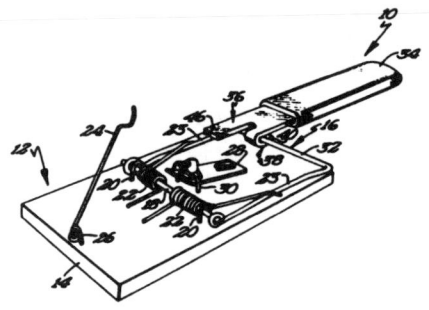

1. Device for assisting the setting of a trap, with the trap
including a striker including a cross bar interconnected to an
arm extending in a nonlinear manner from the cross bar, with
the cross bar being movable and biased toward an entrapment
member, comprising, in combination: an elongated handle; and
means attached to the handle for moving the cross bar against
the bias away from the entrapment member comprising a hook
for removable engagement with the cross bar of the striker of
the trap and a first finger for removable engagement with the
arm of the striker of the trap, with the first finger extending
longitudinally from the handle and including a free end, with
the hook formed by a second finger separate from the first
finger and extending from the handle generally parallel to and
spaced from the first finger and formed into a generally U-
shape, with the hook having a free end extending in a direction
opposite to the free end of the first finger, with the first finger
and the hook being nonmovable relative to each other.

---

to cover unobvious applications, such as control and sequence
of administering the ether.[18] Another classic example clarifies
this distinction further. The fact that food could be frozen was
a scientific fact long accepted before Birdseye became the king
of frozen foods. But Birdseye claimed an unobvious application

of this old principle with amazing results. The basic concept behind this patent (long since expired) was a method of freezing food fast—so fast in fact, that the water in the food had no time to collect into needles of ice which would rupture the food cells.[19]

Multiple Claims

Even the simplest invention will have more than one claim. For example, one of the inventions that my partner, Craig Herrington, was instrumental in getting to market was the Wedgee, an anti-lip sleeve for eyeglasses. The patent for this simple device (which continues to make him money to this day) consisted of six claims.

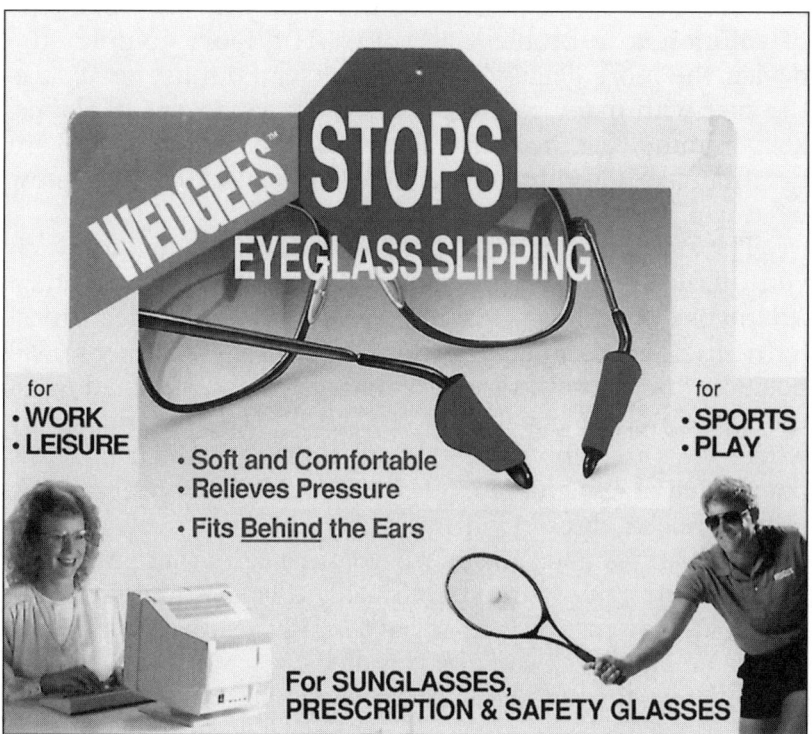

Your invention, then, is likely to have many claims. After the first broad claim, you may make additional claims of diminishing breadth and greater depth, going down the line until you have reached a point where you are claiming practically your entire specification. For example, a claim on a "protective covering" is very broad; the less it says, the broader it is. On the other hand, a claim that covers every brick, shingle, electrical outlet, and water spigot is very deep, narrow, and specific. In such a narrow claim, you can claim minor elements by the simple fact of mentioning them.

However, you cannot make a particular claim for your invention if it has already been claimed (i.e., there is prior art on file with the PTO). Each claim must describe a *new aspect or function* of the device. Thus, you will need to do a preliminary search to locate inventions that deal with the same type of solution to a problem as yours. The more complex the device, the more detailed your patent search must be. If, as is the case with many patents, yours includes dozens of claims, keep in mind that the search fees could escalate quickly. (I am not trying to discourage you here, merely letting you know what you're up against.)

The key to getting a patent approved is the extent to which you utilize the vocabulary contained in the patents you turn up in your preliminary search. By using the same mechanical terms to describe similar portions of your invention, you will position your invention in the patent examiner's channel of thought, in effect enabling to him to see quickly and clearly where your invention differs from those on which previous patents have been granted.[20] If your search turns up something that correlates directly with one of your claims, you can still use that part for your invention; you just can't claim it in your patent. If the patent is more than 20 years old, you can use the technology as part of your invention. Remember, an old idea can be resold; it just can't be patented.

One final tip: the successful drawing of claims also involves, to a certain extent, the use of "nice" language. Take the choice of whether to use "rigid" or "nonflexible" in describing a part. "Rigid" is a positive term, whereas "nonflexible" is negative. As a general rule, direct, positive, and constructive words put the reader in a more cooperative frame of mind.

NOTES

1. Aubrey D. McFadyen. "Have You Fully Worked Out Your Invention?," *Science and Mechanics* (September 1962): 33.
2. Ibid., 35.
3. Ibid., 31.
4. Reprinted and revised from *Design or Mechanical Patent?* by Alan Montague, 101.
5. Ibid., 101.
6. Ibid., 102.
7. Ibid., 103.
8. Art Youngquist, "Which Should You Do—Patent or Copyright?" Inventor's Handbook 1962 Edition, *Science and Mechanics,* 108.
9. Ibid.
10. McFadyen, "Have You Fully Worked Out Your Invention?," 33.
11. Youngquist, The Inventor's Handbook 1962 Edition, *Science and Mechanics*, 50–85.
12. McFadyen, "Have You Fully Worked Out Your Invention?," 33.
13. Youngquist, The Inventor's Handbook 1962 Edition, *Science and Mechanics*, 50–85.
14. McFadyen, "Have You Fully Worked Out Your Invention?," 33.
15. Youngquist, The Inventor's Handbook 1962 Edition, *Science and Mechanics*, 50–85.
16. Ibid.
17. McFadyen. "Have You Fully Worked Out Your Invention?," 33.
18. Ibid.
19. Ibid.
20. Youngquist, The Inventor's Handbook 1962 Edition, *Science and Mechanics*, 50–85.

CHAPTER 6

License Your Invention

So why is it that some inventors become rich and others don't? Almost 99 percent of all patented products lose money! It has to be a good idea to be issued a patent, right? So why is it that only 1 percent of all "good" patented ideas makes money? Whether you get a patent or not, having the knowledge that will allow your invention to make money for you is the key. If it's a good idea, you can sell it.

How does the small percentage of successful products get to market? The core answer to this question is *licensing*. The ability to license your invention is the main—if not the only—reason for patenting your idea. Estimates vary, but based on my personal experience and observation, I would say that 85 percent of all products are licensed to a second or third party—the second party being a manufacturer and the third party being an agent or broker. I will explain the advantages and disadvantages of both shortly, but first let's make sure you've covered all your bases.

In order for you or anyone to produce and market your product, some things need to be in order. First, you must

have taken the important step of building a working prototype of your product. Even though people can understand
drawings and text about a product, there is nothing like a 3-
D product to answer everything from dumb to profound
questions that arise. Second, you need to have tested your
product to ensure it works as you intended. In many cases,
product performance is not as important if the bugs can be
worked out before production. The main idea is that your
product can work and that this is obvious to all who see it.
Unless your idea is something that will change the path of
mankind, it must be something that can be produced using
standard manufacturing processes. It must also be able to be
produced inexpensively so it can turn a profit. Among others,
these are the key components that will sell your idea to a sec-
ond- or third-party licensee.

Finally, you must have protected your product by at least
filing a disclosure document with the PTO.

DDP VS. PATENT

As mentioned in Chapter 1, the PTO's Disclosure
Document Program provides for a two-year limited protection of an inventor's ideas for a small filing fee of $25. In the
past, I have used a disclosure document template (like the one
provided in Appendix II), which is accepted by the PTO. Be
sure to fill out completely in ink, date it, and have two witnesses sign it. I usually have each page notarized as well.
(Most banks provide this service to their customers free of
charge.) You can then file it for safekeeping or send a copy to
the PTO. Sometimes it's good practice to do both, since this
seems to double your chances for proof of conception. It is
well worth the $25 filing fee.

It's also a good idea to send a self-address postcard along
with your disclosure document. See sample below. This will
ensure that the PTO received your submission.

Commissioner of Patents & Trademarks
Document Disclosure Program (Box DD)
Washington, DC 20231

Your Name
Your Address
City, State Zipcode

Received today: Document Disclosure Filing Form.

Check#_____in the amount of $25.00 for filing fee.
Self Addressed stamped envelope.

Duplicate sets of:
Document Disclosure Filing Forms having:

_____pages of written specifications and _____pages
of drawings.

It's important to understand that the disclosure document is not a legal patent, but rather a means to prove that you have dated material pertaining to your invention—conceptual priority of sorts. Nasset offers the following word of caution:

Many inventors believe that when they send in their document disclosure to the Disclosure Document Program (DDP) in Washington, D.C., they have patent-pending status. This simply is not true. The only time an inventor has patent-pending status is after patent application has been properly filed and recorded. To falsely claim patent pending is illegal and punishable by a fine up to $500.[1]

If, on the other hand, you apply for a patent and pay the basic filing fee ($380 for Utility, $155 for Design, and $240 for a Plant patent in 1999 dollars), you have two years in which you can legally claim "patent pending" status for your invention. After that you must complete the patent process or lose your rights. This entails submitting the completed (or revised) patent application with the remainder of the filing fee within two years, or you will lose your right to patent that particular invention. Again, this law is designed to prevent someone from monopolizing an idea by extending the "patent pending" status indefinitely and avoiding paying the full required patent fee.

FINDING A BUYER

So let's assume that you have completed your documentation, signing and notarizing each page. The papers are triplicated and filed in safe places and/or you have filed your patent or disclosure document with the PTO. Your prototype works with spell-binding charm. Now you are ready to sell your precious baby to the highest bidder. There are two ways to do this. First, you could go door-to-door to various manufacturers and try to sell the product yourself. Second, you could approach a marketing agency and, for a cut, let it do all the work required to develop the sales potential of your invention.

Seeking Manufacturing Buyers Yourself

Approaching manufacturers directly is a viable path but a tricky one, as it requires your placing your trust in a stranger. If your product falls into a small niche, this may be the way to go (with discretion). Because all companies want an edge over their competition, they are open to outside ideas. Most companies, large and small, have standard policies and procedures for acquiring new products (consisting of nondisclosure agreements, committee meetings, and so forth). Most companies will go to extremes to protect your idea if approached properly, simply because, in this sue-crazy society, they do not want the reputation of idea theft. Because they want both parties to feel comfortable in the exchange of information and ideas, they will most likely have you fill out a number of forms that protect you and them. *If anything you're asked to sign or disclose appears questionable, have an attorney look it over before you proceed.*

In most cases you will need to have several meetings with a prospective manufacturer. If the manufacturer is far from your home base, this can be time consuming and expensive, and there is no guarantee that your product will be accepted. If you have more than a few companies to approach, it can become a full-time project.

The biggest advantage to approaching a manufacturer directly is that you are directly in control of your product, and if you have good people and business skills, negotiating a better deal for yourself is possible. One thing to remember is that you may not know how their business works, so it is a good idea to do some research in their market area, starting at your local library.

Sealing the Deal

You have found that perfect buyer. He is interested in your invention and quietly alludes to the possibility that he wants to make a deal. Be careful! Do not be deluded by inflated views of your own genius. Many inventors fall into the trap that because they want something, "everybody" will want it. But the fact remains that 9 out of 10 new products prove failures because the public won't buy them. This trap is very seductive. It can even pull in a whole panel of experts consisting of the inventor, patent broker, manufacturer, lawyers, designers, engineers, tool and die makers, ad agencies, package designers, sales managers, jobbers, wholesalers, salesmen, and retailers. If all of these people can be fooled 9 times out of 10, it's easy to see how the individual inventor without marketing experience can easily misjudge public need for something he feels is new and useful.[2]

Take, for example, my 1974 model of the 2-cubic-inch, solid state color organ that connects directly to one's speakers without batteries or AC. Craig and I tried marketing this gem, which turned out to be turd. It looked beautiful while the music played on most units but created distortion problems with the newer, more powerful speakers coming onto the market and sometimes blew out the LEDs. This meant an increase in the cost and size of the unit.

We all tend to overevaluate our own inventions and may feel compelled to hold out for the highest bidder. Consider the true story of a man who tried to sell his invention for only $5,000 because his patent had only one year to run. Investigation revealed that when he was granted his patent in about 1920, he was offered $1 million for it. At the time, he held out for $1.5 million. In 1924, he was offered $750,000 but insisted on $1 million. In 1926, he was offered $500,000 but held out for $750,000. Before he was through, he got down to offering the patent for $25—enough to cover his postage stamps.[3]

If you get a good offer, take it. Don't demand a million marbles for a $10 idea. In the mind of the manufacturer who is buying your idea, that's all he is buying: an idea. He has probably had scores of people, including friends and family, pelting him with new products to sell, and he is already making money, so why should he go very far out of his way for your invention? You have your illusion about your invention. He has lived through several disillusions, some of which have almost certainly cost him money. And to put your idea into production will cost him again initially; whether the risk will pay off in the end remains to be seen.

A Check Would Be Fine, Thank You

In the past, many inventors hesitated to sell patents outright because of the big bite the IRS takes from the money, which is treated as "income." For this reason, royalty payments were often preferred, in that they spread the money out, helping to keep the inventor in a lower tax bracket.

However, since the early 1960s, changes in the law allow you to treat an outright cash sale as a capital gain, so it is generally wisest to sell the rights in one lump sum. (Consult with your CPA regarding what is best in your particular situation.) First of all, it is sometimes very difficult to collect royalties. Second, four out of five new products that do make it to market for a while are said to be failures. Your chances are four to one against there ever being any royalties to collect. Therefore, you would be better off to take the money and run. If you hit upon something in oil cracking, superconduction, or supermaterials, you may be able to sell out for a huge sum.

However, in most cases, because of marketing risks, manufacturers will not pay big money for a patent on a consumer item or small tool. In this case a royalty set-up might be the best, or the only, deal you can get. Sometimes the sale of a patent involves a cash payment on a put-up-or-shut-up basis for a tool that a certain manufacturer wants to take off the market. This is known as "preemptive" buying. The purpose is not to deprive the public of a good invention, but merely to keep a competitor from getting something to sell.[4]

If the invention promises to have far-reaching effects on the economy (such as cheaper energy or faster space travel), then it would be well advised to go for a down payment of a determinate amount of cash (within a reasonable range of about 10 percent of the projected gross annual sales five years from now) plus a yearly royalty of up to 5 percent, but no less than 1.5 percent, depending upon the possible impact of the invention. The latter figure may seem like very little, but consider that 1.5 percent of just $1 million in gross annual sales is $15,000 per year.

In general, most inventions draw the customary royalty of 5 percent of the wholesale price, but few contracts are etched in stone.

Give It to Me in Writing!

In any case, get it in writing! It's imperative to have a contract both parties can agree upon and sign. Insist on a minimum annual royalty of about $1,000 to $1,500—and hedge this with the provision that you can refuse such payment and take the patent back unless the money is actual royalty proceeds from regular sales through normal distribution channels. In other words, don't give the manufacturer the right to keep your patent on the shelf for a token payment of $1,000 a year.[5] Some companies may buy your invention and shelve it until they have more floor space or manufacturing capabilities. A delay in marketing may mean nothing to a manufacturer with a diversified line, but it can be fatal to the inventor whose eggs are all in one basket.[6] You can also negotiate a guaranteed minimum sales over time.

Be sure to have your contract drawn up by someone who is skilled in such documentation.

In the back of this book (see Appendix VIII) are a few examples of contract forms. These are *examples only*; they are not to be used directly without first seeking legal advice. (The authors and publisher take no legal responsibility for the use of these or other forms supplied with this book. These are merely bare-bones examples to give readers a basic idea of

standard formatting of such forms.) These forms are commonly referred to as "boilerplates"; they have all the standard elements of legally binding contracts and yet can be changed to
meet your specific needs. Computer programs are also available through your local computer supply outlets and through
free programs available on the 'net (see www.Legaldoc.com for
additional legal forms).

Using Agents and Brokers to Secure Licensees

Unless you are a veteran in dealing with the business
mentality, product agents and brokers are the way to go in
securing licensees to produce and market your product.

(Product brokers are *not* to be confused with idea or patent
brokering services, which are notorious for ripping people off.)
These are middlemen who will go out and aggressively sell
your invention for a percentage, and sometimes a direct fee up
front as well. (Obviously, a percentage deal for the broker is
ideal, in that it is initially much easier on the inventor's wallet.)
This is how the majority of products gets to market.

Unfortunately, sometimes it is difficult to determine which
agents are best, not to mention legitimate. As a general rule, if
their promises or claims sound to good to be true, they probably are. Beware of people that make claims and can't back
them up. Having had bad experiences with invention brokers
in the past, I will quote directly from Richard C. Levy's book
review, which was posted on Amazon.com:

> Under the guise of providing invention
> marketing services, the dreams of an estimated
> 25,000 inventors are dashed annually by non-
> performing, para-creative slugs who prey on
> inexperienced and needy inventors. The mar-
> keters praise every concept, no matter its
> merit, charge the inventor fees ranging from
> hundreds to thousands of dollars, and do little

> or nothing to commercialize the invention.
> Seventeen (17) states have laws on their books
> that protect inventors, and The Inventor
> Protection Act of 1995, sponsored by U.S. Sen.
> Joseph Lieberman (D-CT), is still in commit-
> tee (summer 1996). In addition to learning
> how to license your invention, my book tells
> you how to avoid rip-off artists (don't ever pay
> an agent a dime to represent your invention)
> and save money with lawyers, patent searchers
> . . . , etc. Remember, information is power.
> Celebrate failure. And Never give up!

It pays to be cautious. That is not to say, however, that there are no good agents out there. In our Internet searches, we have come across a few reputable invention brokers, one in particular is the W.T. Bradley & Sons Enterprises, Inc. Being an inventor himself, Bradley appears to be an honest businessman who has a good marketing plan. Check out The Inventor's Mill-Shop yourself on the Web at www.wtbradley.com.

Good agents and brokers are up front and will give you a realistic view and opinion of your product, what they can do, and who they will approach. Following are a few questions that offer some general guidelines:

- What products they have placed with manufacturers in the past?
- Can they provide references from inventors and manufacturers?
- Do they have previous experience with your type of product in its market?
- Can they refer you to someone else if they cannot help you?
- Are their nondisclosure agreements binding to both parties?
- What are their fees, and what are you getting for it? (In

most cases, agents and brokers to not demand money up front. If they do, find out why, because most get their fees after a deal has been made.)

The list can go on and on, but these are the main things to keep in mind.

Licensing is a whole industry in itself. We are merely touching on the basics here, so you can make some solid decisions on how to promote your product. There are many good books on this subject, and we strongly recommend that you invest in a few so that you are armed with proper knowledge concerning this very important field. In most larger libraries you will find trade directories, including the *Thomas Register* and others, on manufacturing representative organizations. Another excellent resource is the Manufacturers' Agents National Association (MANA) Web site at www.manaonline.com. MANA's online directory provides contact information for tens of thousands of agent rep companies and organizations worldwide for any type of product you can think of.

NOTES

1. William J. Nasset, *This Lit'l Inventor Went to the Patent Office* (Visit his Web site at www.inventorworld.com/ftm-phase1.htm.)
2. Alan Montague, "Putting the Sales Tag on Your Invention," *Science and Mechanics* (September 1962): 84.
3. Ibid.
4. Ibid., 85.
5. Ibid., 84.
6. Alan Montague, "What the Patent Law Means to the Inventor," *Science and Mechanics* (September 1962): 28.

CHAPTER 7

Be Your Own Boss

In the previous chapters, we have presented you with the information you need to protect your product officially should you decide to seek investors or license your product to a second or third party. There is one other route, however, and that is to go it alone: don't even bother with the PTO or third parties; simply document everything yourself and go straight to market with it. Today, I bought a brand new Lufkin tape measure with a neat new thumb toggle-lock. It's a high-quality tool for a fair price. But nowhere on this product could I find the words "Patented" or "Patent Pending."

In many ways, you are really better off not patenting the invention or filing a disclosure document at all. In fact, according to some estimates, only 1 percent of the products on the market today are patented, so apparently there are a lot of people out there who take this viewpoint. Why? Following are a few good reasons:

1. *First one to the marketplace, not the U.S. Patent and Trademark Office, wins.* There are many out there who are willing to break your contract with the government: once you have patented your invention, others now have access to your information and many are ready, willing, and able to steal and sell your invention to a big corporation. Even if you were to take this corporation to the courts, it would be years of heartache and many dollars later before a decision would be handed down. Meanwhile, the big company is making millions of dollars from your work. The main disadvantage of filing a disclosure document is that if you fail to file for a full patent by the end of the two-year period, the device is up for legal grabs. Also, there is a little-known fact that if you publish a work with patent drawings or description, you have one year to begin the patent process, or it can also become public domain.

2. *The U.S. Patent and Trademark Office is not on your side.* There are many documented cases where an inventor was not granted a patent because of trivial prior art issues. Some inventors have even lost their rights for a patent: more than 80 percent of all patent claims brought before the courts are overturned or held invalid.[1] Protection of your invention is not in the PTO's interest. This bureaucracy has become a business with its own interests in mind, operating under the graces of the U.S. government, and has fallen into the pit of corporate greed. Having done a fair amount of research for patents over the years, I have noticed how sloppy the drawings have become. And in most of the recent cases where a patent was issued, the devices were not feasible at all. The PTO does not guarantee the validity of a patent! It will accept almost any concept as long as that cashier's check is in the envelope. This is how greedy this governmental agency has become. It has profited by perpetuating the myth that an inventor needs to patent his invention in order to protect it. This may have been true in the past, when the court system

set up for hearing patent infringements had smaller case loads and actually proved to be a deterrent to such theft. In recent years our government has done little to hasten prosecution of intellectual theft—especially by companies in foreign countries.

Unfortunately, many overseas firms are ripping off hard-earned ideas on the global market these days. Patents are no guarantee of ownership. In fact, if you were to take someone to court for patent infringement, your case may not stand up in court. In fact, in many cases, getting a patent can cause you to lose your invention to a larger corporation—after spending years and a great deal of money in court! Even if you have a solid case, you could still lose out. Take the following, for example:

> Robert W. Kearns, the man who invented the intermittent windshield wiper, claimed that auto makers stole his patent and made hundreds of millions of dollars. He had been in the courts for twenty years and was finally awarded $21 million from Chrysler. Sounds great, huh? Just listen. Legal fees got most of it and Kearns says he's still broke and the lawsuits have consumed him. His wife left him and he was once committed to a psychiatric hospital. So much for protection of a patent![2]

This is not to say that you should not patent your invention. But believe it or not, there are many unscrupulous companies out there with teams of researchers scouring the patent files for new ideas to steal. Patents are easy to circumvent: change a thing here, modify a thing there, and presto—Digi-Scam is selling your idea!

3. *The expense is not worth the effort.* Patents are not self-sustaining; they have to be maintained. A fee must be paid to the

PTO three times within the 20-year protection period, and this can get expensive. The PTO wants to be sure that you are sincere about marketing your invention! Your DDP application is good for only two years, after which, if you have not filed for a full patent, you lose the rights to patent said invention. Yes, it is rewarding to receive a long, hard, sought-after patent. And hanging from the wall, it would impress your friends. On the other hand, you could be making millions of dollars instead. Do bear in mind, however, that according to patent law, if you cross state lines to advertise or sell a prototype or production unit before you begin the patenting process, you lose the right to patent it. Once the invention has been published, it enters the realm of public domain and is up for grabs. Personally, I would try a *national* advertising campaign. Of course, if it were a good concept and someone could prove that you did it, there is always the danger that he or she might contest it in court so that they could steal and sell your product. But in most cases, you could get away with a national advertising test run. If it succeeds, you could make millions before anyone takes notice of what you have done, and then who needs a patent anyway? (If you still decide to patent your idea, use caution. Advertise the product first, *in state*, and see what kind of response you get—*then* produce it! If your response is low, send a letter to the few who ordered explaining—with a refund—that the product is not available for one reason or another. You will have saved yourself thousands of dollars.

4. *You can protect it yourself.* With your time and money on the line, you have to decide whether a patent would actually increase the value of your invention. All things considered, maybe it would be better to spend your time and money on determining how to make it, package it, and market it. One thing is for certain: *you must get it to the marketplace fast!* If it turns out that your product sells, you'll probably still have at least a year to apply for a patent. With the speed at which

information can be conveyed today, you might do well to simply prepare and date the patent, get two witnesses to notarize it, and file it for your own safekeeping instead of sending it in to the Patent Office right away. Should you get wind of another invention such as yours, you are prepared to submit your invention to the PTO to be reviewed immediately if you decide you need to do so. (Remember, this gives you the chance to see your competitor's work as well.)

If it turns out that your product doesn't sell, look at the money you have saved! Consider the sage advice of Dr. Vernon Brabham: "The ultimate goal of having an idea for a new product is to make money, and if getting a patent is the proper tool to achieve that goal then go for it, but don't do it just because someone says you should. If getting a patent is not the proper tool, then spend money and time getting your product out there fast, first, and profitably!"[3]

In general, it would appear that you are better off documenting and safekeeping your invention yourself and then striking out on your own and marketing it yourself. In reality, this is probably the safest, fastest, and most profitable way to get your product out there. (If it seems to you that we are contradicting what we said earlier in this book, this is merely because all aspects of the patent process must be presented to you before you can make an informed decision.) However, marketing your product yourself does involve more work, and the stress factor definitely goes up. Inventing a product is one thing, but marketing is another whole can of worms. Most inventors would rather not have to deal directly with the public. However, the potential of the Internet is just plain fantastic.

THE BEAUTY OF THE INTERNET

If you can create a fetching and effective Web site, you'll likely do very well. Some inventors are making millions of dollars on the Internet—without leaving home—and working just a few hours a day! You can too! Thanks to the 'Net, it has become much easier to get the word out and sell to consumers directly. All you need to do is follow a few simple steps:

1. Get online
2. Set up your Web site
3. List your site with the appropriate "search engines"
4. Set up a credit card payment system
5. E-mail possible buyers
6. Stay on top of e-mail orders

The first step is to get your URL, or Web site address. Your address will begin with www, which stands for Worldwide Web. The .com suffix indicates that yours is a commercial site (as opposed to .org, for nonprofit, or .gov, for government, and so forth). So, your new address would be something like www.jigglemaroo.com, for example.

Pick your URL carefully. It will be your customer's first hint of what you are about. Unless you are an actor, avoid personal names, which reflect high self-image and lack of creativity. Come up with a company name that reveals what you are selling and is easy to remember. The shorter the name, the more memorable it will be to the customer. Another tactic is to use a company name that will prompt an image in the reader's mind. We remember pictures and scenes better than concepts that require multiple synapses or circuits in the brain. For example, www.amazon.com is more memorable than www.lotsabooks.com.

Once you have an address, you'll want to get online as soon as possible. If computer programming, graphic design, and writing are not your strong points, consider having professionals set up your Web site. Review it carefully, since it is the primary vehicle you will be using to present your invention to the whole world. Most Web sites have more than one page, but unless you are selling a variety of products, you are better off with a single "teaser" page and two or three more detailed pages—none of which will completely give away your invention.

There are many good Web designers out there working for Internet providers and services who not only do good work but wouldn't mind doing a little moonlighting under the table for a fee. Get online and look at what is out there in your field. Look at sites that grab your attention and apply some of the same design elements to your own site. Avoid making your page too fancy or complex, which increases the time it takes to download your site. Many 'Net shoppers tend to be impatient and will not wait if your site takes too long to download. This is because there is just so much information on the 'Net to discover! The idea is to keep it simple, enticing, and memorable.

To ensure that your target customers can find your product easily and quickly, you'll want to register your site with the major search engines. Search engines are powerful programs that scour the Internet for specific items or information about a product or service utilizing key words entered by the person initiating the search. For example, say that I am selling a new type of lawnmower that I invented. I would want to use keywords such as grass, lawn, yard, patio, deck, fence, weed, and so on in my site to increase the likelihood that the search engines will pull up my site when someone uses any of these words in initiating a search. I also want to use very accurate descriptive words in the title of my site (which is different from the address and is embedded in the programming of a Web site). These tactics will also improve the odds that a potential customer using any of the search engines you are registered with will pull up your company address. And then, with the click of a mouse, these customers can visit your site and learn everything you want them to know, including how to order your product.

If you decide to manufacture and sell your own invention,

you need to have the ability to make the sale quick. Credit card payment is the safest and fastest way to conduct business on the Internet. Almost 95 percent of all Internet sales are made because of the immediate availability of the product. In short, the customer should be able to purchase your product online in one "quick-click" transaction. If the consumer has too much time to think about it, then he will probably change his mind, and you've lost the sale.

The next step is to e-mail your growing list of customers, as well lists of potential customers that you are able to compile. For the start-up company today, the ability to do direct marketing via electronic mail is the best thing since the invention of the telephone. I remember back in the 1970s, when direct marketing meant buying targeted mailing lists of people who were likely to be interested in your product based on past buying history, and then, having stuffed thousands of envelopes, hauling them down to the post office to have them metered. If you were very lucky, for all of your trouble you might get two or three orders out of a hundred pieces mailed.

Finally, stay on top of it all. It's crucial that you respond to e-mail requests for products and information and fulfill online orders immediately. In today's environment, if the consumer cannot get it fast, then he will not stay your customer for long. In the early days of mail order marketing, it was easier to sell your product because time was on your side. Usually, you had three to six weeks to deliver the product to the customer. You could spend 70 percent of your energy on running your business and 30 percent of it on responding to your customers. Today, with the lightning-quick rate of Internet commerce forcing us to be more competitive, those figures have flip-flopped.

Fast, reliable customer service is more critical now than ever. If a Web-based company does not deliver what its site promises, the customer's bad experience will be amplified to thousands of others via newsgroups, discussion groups, and "consumer experiences" chats. The flipside of this is that customer who has a good experience can help to advertise your company and product line in the same way.

There are many good books on effective Internet marketing techniques and strategies, so for more detailed information on implementing the steps outlined in this chapter, get online and check them out. For some helpful tips and free advice on setting up your Web site, visit W.T. Bradley's Inventor's Mill-Shop at www.wtbradley.com.

To manufacture and market your own invention is a monumental task for any one person to undertake. If you are a production or factory worker or a professional tradesman, unwise in the ways of the business jungle, you may want to sell out for a little cash and a reasonable royalty, safe in knowing that you haven't lost anything—and may collect a lot. But if you are an experienced businessman, willing and able to forge ahead with all of the complicated facets of managing a company, then go ahead—give it a shot. If you can easily make your product in your basement in quantities of 100 in just a few hours a day, then you will probably make money marketing it yourself on the Internet. If you wish to take on the task of online marketing, your profits can be great.

Best luck on your new career!

NOTES

1. Dr. Vernon Brabham, "The Awful Truth about Patents" (www.bizine.com/patents.htm, 1997): 1. (Contact him at vbrabham@mindspring.com.)
2. Ed Justice, *WorldWide E* (April 1998), www.AEPublishing.com.
3. Brabham, "The Awful Truth," 2.

APPENDIX I

FORM PS-9a: Nondisclosure Confidentiality Agreement

NONDISCLOSURE CONFIDENTIALITY AGREEMENT
THIS IS A CONFIDENTIALITY AGREEMENT, between

_____OWNER
and _____RECIPIENT.

The parties have agreed to the following terms governing the confidentiality of certain proprietary information one party ("Owner") may disclose to the other party ("Recipient");

DEFINITIONS

A. For the purpose of this Confidentiality Agreement, "Confidential Information" means all information in whatever form transmitted relating to the past, present or future business affairs, including without limitation, research, development, or business plans, operations or systems, of Owner or other or another party whose information Owner has in its possession under obligations of confidentiality, which (i) is disclosed by Owner or its affiliates, bearing an appropriate legend indicating its confidential or proprietary nature or otherwise disclosed in a manner consistent with its confidential or proprietary nature or (ii) is produced or developed during the working relationship between the parties and which would, if disclosed to competitors of either party, give or increase such competitors' advantage over the party or diminish that party's advantage of its' competitors.

Confidential information shall not include any information of an Owner that: (i) is already known to Recipient at the time of its disclosure; (ii) is or becomes publicly known through no wrongful act of Recipient; (iii) is communicated to a third party with express written consent of the Owner; (iv) has been independently developed by Recipient prior to receipt of any information disclosed by the Owner or, (v) is lawfully required to be disclosed to any governmental agency or is otherwise required to be disclosed by law, provided that before making such disclosure the Recipient shall immediately give the Owner written notice and an adequate opportunity to raise an objection or to take action to assure confidential handling of such information.

B.   The term "affiliate" shall mean any person or entity controlling, controlled by or under common control with a party.

TERM

For a period of three (3) years from the date of disclosure, to Recipient, Recipient shall not disclose any Confidential Information it receives from Owner to any person or entity except employees of Recipient and its affiliates who have a need to know and who have been informed of Recipient's obligations under this Confidentiality Agreement. Recipient shall use not less than the same degree of care to avoid disclosure of such Confidential information as Recipient uses for its own confidential information of like importance.

OWNERSHIP

All Confidential Information disclosed by Owner to Recipient under this Confidentiality Agreement in tangible form (including, without limitation, information incorporated in computer software or held in electronic storage media) shall be and remain property of the owner. As such Confidential Information shall be returned to Owner promptly upon written request and shall not be retained in any form by Recipient. The Rights and obligations of the parties under this Confidentiality Agreement shall survive any such return of Confidential Information.

TERMINATION

Either party may terminate this Confidentiality agreement by written notice to the other. However, in the event of termination, all rights and obligations under this agreement shall survive with respect to Confidential Information disclosed prior to such termination.

LIABILITY

A. The Parties agree that in the event of a breach or threatened breach of the terms of this Confidentiality Agreement, Owner shall be entitled to an injunction prohibiting any such breach. Any such relief shall be in addition to and not in lieu of any other legal or equitable relief including money damages. The parties acknowledge that Confidential Information is valuable and unique and that disclosure in breach of this Confidentiality Agreement will result in irreparable injury to Owner.

B. Neither party shall in any way or in any form disclose, publicize or advertise in any manner the discussions that are the reasons for this Confidentiality Agreement or the discussions or negotiations covered by the Confidentiality Agreement without the prior written consent of the other party.

COMPLETE AGREEMENT, AMENDMENTS, GOVERNING LAW

This Confidentiality Agreement: (i) is the complete agreement of the parties concerning this subject matter and supersedes any prior such agreements; (ii) may not be amended except in writing signed by both parties; (iii) shall be governed by and in accordance with the laws of the State of _____ without its choice of law provisions; and (v) is executed by authorized representatives of each party.

OWNER

By:_____Date_____

Print Name:

Print Title:

Product Name:

RECIPIENT

By:_____Date_____

Print Name:

Print Title:

APPENDIX II

··

FORM PS-1b:
Sample Disclosure Document

Title: Inertial Propulsion Demonstration Device (Model E6)

Purpose: To demonstrate the rectification of a mechanical oscillator to produce unidirectional impulses for propulsion from the centrifugal force of eccentric rotors. The purpose of the system is to increase the critical action time (CAT) on the positive phase of the cycle. This creates a biased mechanical oscillator in the positive direction—thus allowing the rotor time to impulse with centrifugal force across the forward side of the mainframe, propelling it forward.

Description: Based on the principle of the Dean Drive (Pat. #2,886,976), an electric motor supplies angular torque to a single eccentric rotor. In this modification, the single rotor is mounted on a nylon cam which rides a roller/lever microswitch. As the rotor spins, it propels the motor and its carriage forward and backward with equal force. This results in a balanced oscillation of the carriage with its motor, rotor and commutator system on the spring-loaded rods. The springs supply a central reference for the carriage. The two rods convert the 360° motion of the rotor into a bi-directional momentum of the carriage. In this fashion, major side forces are canceled out and the carriage complex swings equally—to and fro—under the centrifugal force of the rotor.

With the solenoid circuit switched on, the cam activates the microswitch at the proper time each cycle—firing the solenoid—and the carriage complex is advanced in the positive direction before the rotor has time to drive it there. With the carriage already forward, the resultant force from the rotor is imparted upon the mainframe. If the solenoid is properly set, there should be no impact—just a quiet, clean-sweeping centrifugal force from the rotor itself.

The mainframe is mounted on four, free-wheeling roller shaft bearings acting as low-frictional wheels. The wrinkled (or dimpled), high-gloss black epoxy finish on the rotor is for reduced air resistance, durability and visual contrast.

A fast recovery diode, IN5819, is electrically across the solenoid coil to dampen back EMP. Without it, the collapsing magnetic field delays the release of the carriage, disrupting the rotor/carriage cycle. It is mounted on a terminal strip for easy replacement and allows access to test the solenoid. A capacitor, .01 uf @ 400vdc, is connected across the microswitch for arc suppression on the switch's contacts, thus increasing the life of the microswitch and preventing contact sticking.

The switch plate holds two miniature toggle switches with a green LED cycle indictor. The motor switch with the tilted ellipse pattern represents the rotor's roto-linear acceleration while the shifter switch with the indicator LED displays a biased sine wave. A current limiting resistor of 400 ohms, 1/8 watt is in series with the LED for a 12 VDC shifter system.

(Since 3- to 6-volt DC solenoids are not readily available, we had to use a 12-volt system for the 12V solenoid shifter. For the convenience of the demonstration, we use this large 12-volt source. With the proper solenoid, we could reduce this battery pack down to a size where it could ride on the mainframe of the engine.)

Though one rotor works with such a small, inefficient system, two rotors provide greater stability and thrust along the plane of oscillation. See additional design claims below (Figure 1) of rotor placement (rotor stacking).

Counter eccentrics supply stability along the plane of oscillation by cancellation forces. These rotors can be arranged as depicted as long as they counter-rotate on this plane.

The plate mounted to the bottom of the motor is the shifter linkage plate

Though one rotor works with such a small, inefficient system, two rotors provide greater stability and thrust along the plane of oscillation. See additional design claims below (Figure 1.) of rotor placement (rotor stacking)

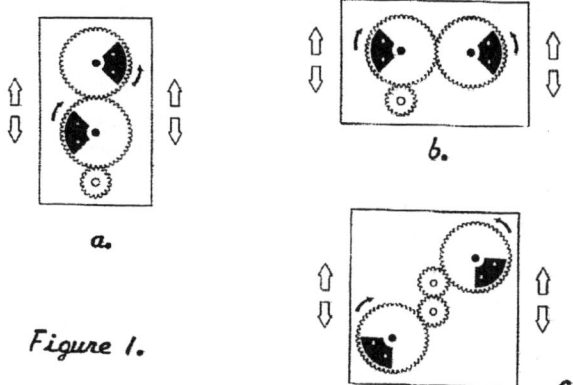

Figure 1.

Counter eccentrics supply stability along the plane of oscillation by cancellation forces. These rotors can be arranged as depicted as long as they counter-rotate on this plane,

The plate mounted to the bottom of the motor - is the shifter linkage plate and mechanically joins the carriage/rotor system to the solenoid plunger. Engine weighs 2 lbs., rotor = 2 oz. total power consumption is 12 watts. Thrust: about 3 to 4 oz.

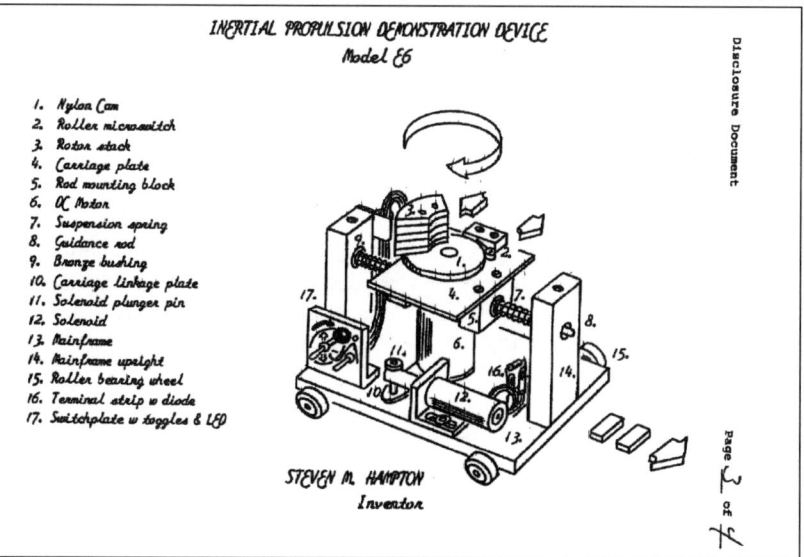

INERTIAL PROPULSION DEMONSTRATION DEVICE
Model E6

1. Nylon Cam
2. Roller microswitch
3. Rotor stack
4. Carriage plate
5. Rod mounting block
6. DC Motor
7. Suspension spring
8. Guidance rod
9. Bronze bushing
10. Carriage linkage plate
11. Solenoid plunger pin
12. Solenoid
13. Mainframe
14. Mainframe upright
15. Roller bearing wheel
16. Terminal strip w diode
17. Switchplate w toggles & LED

STEVEN M. HAMPTON
Inventor

Disclosure Document

and mechanically joins the carriage/rotor system to the solenoid plunger. Engine weighs 2 lbs., rotor = 2 oz. total power consumption is 12 watts. Thrust: about 3 to 4 oz.

The white plastic microswitch cover plate along with the white coating of the carriage plate supplies a contrasting backdrop for the black rotor, allowing the observation of "phasing."

This tilted elliptical pattern is typical of Dean Drive rotors that are in or near thrust mode where the acceleration vector lags the velocity vector by as much as 45 degrees past normal concentric rotation. This totals to a 135-degree phase lag in these eccentric rotors and in fact, does not conform with the classical notion of a mechanical ellipsoid orbit. The rotor drives the frame 45 degrees ahead of itself, another enigma. And to top it all off, shouldn't the carriage (white plate mounted to the spring-loaded rods) be propelled back, 180 degrees away from the rotor's swing? All in all, the system is 225 degrees out of whack with Classical Mechanics.

The unidirectional force emitted by this "toy" is a product of broken symmetry which reflects the problem of dimensional stress caused by gyroscopes. Since more energy is going into the system then what it is ultimately producing in thrust, the net conclusion is that gravity/inertial forces are being emitted on the forward side of the system while running. An experiment is currently being set up to detect this "Inertial radiation."

Conception Information Date and Location: On 14 Aug. 1995 Mr. Thomas Valone, M.A., P.E., requested a demonstration device for inertial propulsion. Device completed and shipped 30 Aug. 1995.

Ramifications: As a method of space drive, this system could revolutionize the transportation industry.

Possible Novel Features: Simplicity of the principle.

Advantages: 1. According to Mr. Valone, this system is five times more efficient than the equal mass of a jet or rocket engine. 2. Safer than conventional reaction/mass engines. 3. As a space drive, this system can be simplified to very few moving parts—unlike the complex mechanisms used to control dangerous jet and rocket engines.

LT-01 DISCLOSURE DOCUMENT

Title:_____

Purpose:_____

Description:

Conception Information Date and Location:

Ramifications:

Possible Novel Features:

Advantages:

INVENTOR   DATED

_____   _____

WITNESSED AND UNDERSTOOD:

_____   ___/_____/_____

_____   ___/_____/___FORM
Title:_____

Advantages   Continued

DISCLOSURE DOCUMENT DRAWINGS   Page of

INVENTOR:_____
Date:____/_____/_____

INVENTION
TITLE:_____

COMMENTS

APPENDIX III

FORM DD:
DDP Request Letter

Date:

Disclosure Document Program (DDP)
Commissioner of Patents and Trademarks
Washington, District of Columbia 20231

Request for Participation in Disclosure Document Program:

Disclosure of _____

Your Name(s)

Entitled: _____

Title of Disclosure

Sir:
Attached is a disclosure of my above-entitled invention (consisting of _____ sheets of written description and _____separate drawings or photos), A $_____ check, a stamped, addressed return envelope, and a duplicate copy of this letter.

The undersigned respectfully request that this disclosure be accepted and retained for two years (or longer if it is later referred to in a paper filed in a patent application) under the Disclosure Document Program and that the enclosed duplicate of this letter be dated stamped, numbered and returned in the envelope also enclosed.

The undersigned understands that (1) this document is neither a patent application nor a substitute for one, (2) its receipt date will not become the effective filing date of a later-filed patent application, (3) it will be retained for two years and then destroyed unless it is referred to in a patent application, (4) this two-year retention period is not a "grace period" during which a patent application can be filed without loss of benefits, (5) in addition to this document, proof of diligence in building and testing the invention, and/or filing a patent application on the invention, may be vital in case of an interference, and other situations, (6) if such building and testing is done, signed, and dated, records of such should additionally be made and these should be witnessed and dated by disinterested individuals (not the PTO), and (7) if any public use or sale of the invention is made in the US, or any publication is made anywhere, no valid patent can be granted on the invention unless a patent application is filed on it within one year of any such public use, sale of publication, regardless, of the filing date of this Disclosure Document.

Very respectfully,

_____   _____   _____
Signature of Inventor   Signature of Joint Inventor   DATE

_____   _____
c/o (print name)   Print Name

_____   _____   _____
Address   Address   DATE

FORM DD

APPENDIX IV
Sample Patent Specification

APPLICATION FOR
UNITED STATES LETTERS PATENT
SPECIFICATION

TO ALL WHOM IT MAY CONCERN:

Be it known that I, John M. Smith, a citizen of the United States of
America and resident of the State of California, having a postal address of
123 Main Boulevard, Anytown, California 9####, have invented a new
and useful "Method of Fabricating Friction Members for Eyeglasses," of
which the following forms the specification.

Method of Fabricating Friction Members for Eyeglasses

<u>Technical Field</u>
The present invention relates to the field of eyeglass holder manufacturing, and in particular to a method of fabricating friction members which
engage the temple arms of a pair of eyeglasses.

Background of the Invention

The present invention arose from an attempt to develop a method of quickly, simply, and inexpensively mass-producing eyeglass temple cushioning and wedging friction members such as those disclosed in U.S. patent No. 4,848,861.

After a protracted period of trial and error compounded by the particular difficulty in finding a suitable adhesive for the finished product, which would produce a secure bond between the opposed surfaces of the friction material, yet remain flexible so as to allow the friction material to flex and stretch without being noticeably restricted by the presence of the adhesives, the method that forms the basis of the present invention was finally settled upon.

This method represents the most direct, efficient, and inexpensive means of fabricating friction members for eyeglasses so far developed and should prove to reduce the ultimate cost passed on to the consuming public to the point that every owner of a pair of eyeglasses should be able to afford an unlimited number of pairs of the finished product.

Summary of the Invention

Briefly stated the method that forms the basis of the present invention comprises the steps of: positioning a generally flat sheet of friction material beneath a cutting unit, followed by the step of at least partially severing the outline of a plurality of the unfolded friction members through the thickness of the friction material; then one side of the partially severed portion of the sheet of friction material is coated with an adhesive and the coated surfaces are brought into bonding contact with one another by folding to adhesively secure the friction member in its finished configuration; then the friction members are stitched together through the exterior surface of the friction member in the vicinity of the adhesively joined interior surfaces of the friction members; and, finally the answered peripheral portions of the friction members are severed to produce the finished product.

Brief Description of the Drawings

These and other objects, advantages, and novel features of the invention will become apparent from the detailed description of the best mode for carrying out the preferred embodiment of the invention which follows, particularly when considered in conjunction with the accompanying drawings, wherein:

Fig. 1 is a perspective view of the sequential steps employed in the method of this invention.

Fig. 2 is a top plan view of one of the friction members after the initial cutting step.

Fig. 3 is a perspective view of one of the friction members subsequent to the adhesive coating step and during the course of the folding step.

Fig. 4 is a top plan view of one of the friction members at the end of the stitching step.

Fig. 5 is a perspective view of the finished friction member.

Best Mode for Carrying Out the Invention

The sequential steps that comprise the method that forms the basis of the present invention are depicted in Fig. 1; wherein, as viewed from top to bottom the viewer is looking at a partial severing step; a bonding and folding step; a stitching step; and a final severing step.

As shown in Fig. 1, a generally rectangular sheet of friction material (10) having an outer covering of fabric (ii) is inserted into a severing device (not shown) such as a die cutting machine or programmable laser cutting machine, wherein the severing device is used to partially sever the outline of a plurality of the friction members (100) in their unfolded flat configuration.

At the very top row of Fig. 1 the outline of the friction members (100) are shown in phantom; and, in the next row and in Fig. 2 the partially severed outlines are depicted; wherein, in the preferred embodiment of the Invention are plurality of discrete unsevered lands (12) of material are disposed at spaced locations around the lower periphery of the outline.

As shown in the next row of Fig. 1 and in Fig. 3, the ends of the upper and lower half (101) (101') on one side of the enlarged ends (103) of the friction member (100) are bonded together such as by the application of adhesives (102) prior to the upper half being folded over on the lower half; or by heat sealing (not shown) after the upper half has been folded over on the lower half; wherein the lower half of the friction member (100) is still connected to the remainder of the sheet (101) by the discrete lands (12).

Turning now to Fig. 4 and the next to the last step in the method involves the stitching of A pattern (15) through the lower portion of the double thickness friction member (100)i wherein, a relatively straight stitch is formed proximate the midpoint of the folded friction member;

and, wherein one end of the stitching forms a downwardly
depending loop on the enlarged end (103) of the friction member (100).

In addition as can be seen in the last row of Fig. 1, the final step in the
method involves another severing operation wherein the discrete lands
(12) of each friction member (100) are severed to completely disengage
the finished friction member (100) from the sheet of friction material (10).

At this juncture it would be advisable to mention the preferred materials
and mechanisms contemplated for use in conjunction with the method
steps of this invention. To begin with, the friction material should preferably be chosen from among a class of materials that includes but is not
limited to rubber, neoprene, Hypalon Silicone, etc. The fabric material
should preferably be chosen from among a class of both natural and
manmade fibers which includes but is not limited to: cotton, rayon,
nylon, Lycra, etc.

With regard to the initial and final severing steps this invention presently
contemplates the use of metal dyes and/or the use of a programmable
laser cutting device to sever the multiple fabric member outlines from the
single sheet of friction material.

Furthermore, with respect to the bonding step, this invention contemplates the use of adhesives and/or heat or pressure bonding to secure the
opposed faces of the friction member together; wherein, in the case of
adhesives the preferred type of adhesive in a latex contact cement adhesive for the flexible nature of the bond that is produced. In addition, the
adhesives can be applied by spraying, coating,* silkscreening, or other
suitable means.

Having thereby described the subject matter of the present invention, it
should be apparent that many substitutions, modifications, and variations of
the invention are possible in light of the above teachings. It is therefore to
be understood that the invention as taught and described herein is only to
be limited to the extent of the breadth and scope of the appended claims.

CLAIMS

1. A method for fabricating friction members having an enlarged end
    and a reduced and for use on eyeglasses wherein the method com-
    prises the following steps:

a) positioning a sheet of friction material relative to a severing device;

b) partially severing the outline of a plurality of the friction members, wherein the partial severing step leaves a plurality of spaced lands between the partially severed friction members and the uncovered portions of the sheet of friction material;

c) folding and bonding the opposed faces on one side of the enlarged and of the folded friction members together; and

d) severing the lands between the friction members and the uncovered portions of the friction material.

2. The method as in claim I further comprising an intermediate stop e) between step c and d; wherein the intermediate step comprises

e) stitching a pattern on the double thickness enlarged ends of the friction member.

3. The method an in claim 2; wherein, the sheet of friction material is provided with a fabric covering.

4. The method as in claim 2; wherein, the bonding of the opposed faces on one side of the enlarged and of the folded friction members in accomplished with adhesives.

5. The method as in claim 4; wherein, the stitching pattern comprises: an elongated generally straight stitch.

6. The method as in claim 5; wherein, the elongated generally straight stitch has one end which forms a downwardly depending loop on the enlarged end of the individual friction members.

Abstract of the Disclosure

A method of fabricating a plurality of friction members (100) for use on the temple arms of eyeglasses, wherein the method comprises the steps of partially severing, folding and bonding, stitching, and then completely severing a plurality of friction members (100) from a sheet of friction material (10) having a fabric covering (11).

APPENDIX V

U.S. Patent Office Price Lists

U.S. PATENT AND TRADEMARK OFFICE
Effective November 10, 1998**

The U.S. Patent and Trademark Office (PTO) has amended its rules of practice in patent cases, Part I of title 37, Code of Federal Regulations, to adjust patent statutory fee amounts to conform to Public Law 105-358. Any fee amount paid on or after November 10, 1998**, must be paid, in the revised amount. For the date of mailing indicated on a proper Certificate of Mailing or Transmission, please refer to *the Official Gazette of the United States Patent and Trademark Office*, dated September 15, 1998, in Volume 1214, which is available on the PTO Web site at www.uspto.gov. The fees subject to reduction for small entities who have established status (37 CFR 1.27) are shown in a separate column. For additional information, contact the PTO General Information Services Division at (703) 308-4357 or (800) PTO-9199.

The effective date for the amendments to the fee amounts in 3 7 CFR 1. 17(r) and (s) is December 8, 1998, the date of publication in the *Federal Register*.

Code	37 CFR	Description	Fee (if applicable)	Patent Filing Fees
101/201	1.16(a)	Basic filing fee – Utility	760.00	380.00
102/202	1.16(b)	Independent claims in excess of three	78.00	39.00
103/203	1.16(c)	Claims in excess of twenty	18.00	9.00
104/204	1.16(d)	Multiple dependent claim	260.00	130.00
105/205	1.16(e)	Surcharge – Late filing fee or oath or declaration	30.00	65.00
106/206	1.16(f)	Design filing fee	310.00	155.00
107/207	1.16(g)	Plant filing fee	480.00	240.00
108/208	1.16(h)	Reissue filing fee	760.00	380.00
109/209	1.16(i)	Reissue independent claims over original patent	78.00	39.00
110/210	1.16(j)	Reissue claims in excess of 20 and over original patent	18.00	9.00
114/214	1.16(k)	Provisional application filing fee	150.00	75.00
127/227	1.16(l)	Surcharge – Late provisional filing fee or cover sheet	50.00	25.00
139	1.17(k)	Non-English specification	130.00	

Patent Issue Fees

Code	37 CFR	Description	Fee (if applicable)	Patent Filing Fees
142/242	1.18(a)	Utility issue fee	1,210.00	605.00
143/243	1.18(b)	Design issue fee	430.00	215.00
144/244	1.18(c)	Plant issue fee	580.00	290.00

Patent Maintenance Fees

Code	37 CFR	Description	Fee (if applicable)	Patent Filing Fees
183/283	1.20(e)	Due at 3.5 years	940.00	470.00
184/284	1.20(f)	Due at 7.5 years	1,900.00	950.00
185/285	1.20(g)	Due at 11.5 years	2,910.00	1,455.00
186/286	1.20(h)	Surcharge – Late payment within 6 months	130.00	65.00
187	1.20(l)(1)	Surcharge after expiration – Late payment is unavoidable	700.00	
188	1.20(l)(2)	Surcharge after expiration – Late payment unintentional	1,640.00	

Miscellaneous Patent Fees

Code	37 CFR	Description	Fee (if applicable)	
111	1.200)(1)	Extension of term of patent	1,120.00	
124	1.200)(2)	Initial application for interim extension (see 37 CFR 1.790)	420.00	
125	1.200)(3)	Subsequent application for interim extension (see 37 CFR 1.790)	220.00	
112	1.17(n)	Requesting publication of SIR – Prior to examiner's action	920.00*	
113	1.17(o)	Requesting publication of SIR – After examiner's action	1,840.00*	

146/246	1.17(r)	For filing a submission after final rejection (see 37 CFR 1. 129(a))	760.00	380.00
149/249	1.17(s)	For each additional invention to be examined (see 37 CFR 1. 129(b))	760.00	380.00
145	1.20(a)	Certificate of correction	100.00	
147	1.20(c)	For filing a request for reexamination	2,520.00	
148/248	1.20(d)	Statutory disclaimer	110.00	55.00

* Reduced by basic filing fee paid.

REMITTANCES FROM FOREIGN COUNTRIES MUST BE PAYABLE AND IMMEDIATELY
NEGOTIABLE IN THE UNITED STATES FOR THE FULL AMOUNT OF THE FEE REQUIRED

Code	37 CFR	Description	Fee (if applicable)	

Patent Extension Fees

Code	37 CFR	Description	Fee	Small Entity
115/215	1.17(a)(1)	Extension for response within first month	110.00	55.00
116/216	1.17(a)(2)	Extension for response within second month	380.00	190.00
117/217	1.17(a)(3)	Extension for response within third month	870.00	435.00
118/218	1.17(a)(4)	Extension for response within fourth month	1,360.00	680.00
128/228	1.17(a)(5)	Extension for response within fifth month	1,850.00	925.00

Patent Appeals/Interference Fees

Code	37 CFR	Description	Fee	Small Entity
119/219	1.17(b)	Notice of appeal	300.00	150.00
120/220	1.17(c)	Filing a brief in support of an appeal	300.00	150.00
121/221	1.17(d)	Request for oral hearing	260.00	130.00

Patent Petition Fees

Code	37 CFR	Description	Fee	Small Entity
122	1.17(h) or (i)	Petitions to the Commissioner, unless otherwise specified		130.00
126	1.17(p)	Submission of an Information Disclosure Statement (ß 1.97(c))		240.00
138	1.17(o)	Petition to institute a public use proceeding		1,510.00
140/240	1.17(l)	Petition to revive unavoidably abandoned application	110.00	55.00
141/241	1.17(m)	Petition to revive unintentionally abandoned application	1,210.00	605.00
123	1.17(q)	Petitions related to provisional applications		50.00

PCT Fees – National Stage

Code	37 CFR	Description	Fee	Small Entity
154/254	1.492(e)	Oath or declaration after twenty or thirty months from priority date	130.00	65.00
156	1.492(f)	English translation after twenty or thirty months from priority date	130.00	
956/957	1.492(a)(1)	IPEA – U.S.	670.00	335.00
958/959	1.492(a)(2)	ISA – U.S.	760.00	380.00
960/961	1.492(a)(3)	PTO not ISA or IPEA	970.00	485.00
962/963	1.492(a)(4)	Claims meet PCT Article 33(l)-(4) – IPEA – U.S.	96.00	48.00
964/965	1.492(b)	Claims – extra independent (over three)	78.00	39.00
966/967	1.492(c)	Claims - extra total (over twenty)	18.00	9.00
968/969	1.492(d)	Claims - multiple dependent	260.00	130.00
970/971	1.492(a)(5)	For filing with EPO or JPO search report	840.00	420.00

PCT Fees – International Stage

Code	37 CFR	Description	Fee	Small Entity
150	1.445(a)(1)	Transmittal fee	240.00	

151	1.445(a)(2)	PCT search fee – no U.S. application	700.00
152	1.445(a)(3)	Supplemental search per additional invention	210.00
153	1.445(a)(2)	PCT search - prior U.S. application	450.00
190	1.482(a)(1)	Preliminary examination fee – ISA was the U.S.	490.00
191	1.482(a)(1)	Preliminary examination fee – ISA not the U.S.	750.00
192	1.482(a)(2)	Additional invention – ISA was the U.S.	140.00
193	1.482(a)(2)	Additional invention – ISA not the U.S	270.00

PCT Fees to WIPO

800	Basic fee (first thirty pages)	455.00*
801	Basic supplemental fee (for each page over thirty)	10.00*
802	International search	1,250.00*
803	Handling fee	162.00*
899	Designation fee per country	105.00*

PCT Fee to EPO

WIPO and EPO fees subject to periodic change due to fluctuations in exchange rate. Refer to the *Official Gazette of the United States Patent and Trademark Office* for current amounts.

REMITTANCES FROM FOREIGN COUNTRIES MUST BE PAYABLE AND IMMEDIATELY NEGOTIABLE IN THE UNITED STATES FOR THE FULL AMOUNT OF THE FEE REQUIRED

Code	37 CFR	Description	Fee Fee
Patent	Service Fees		
561	1.19(a)(1)(i)	Printed copy of patent w/o color, regular service	3.00
562	1.19(a)(1)(ii)	Printed copy of patent w/o color, overnight delivery to PTO box or overnight fax	6.00
563	1.19(a)(1)(iii)	Printed copy of patent w/o color, ordered via expedited mail or fax, exp. service	25.00
564	1.19(a)(2)	Printed copy of plant patent, in color	15.00
565	1.19(a)(3)	Copy of utility patent or SIR, with color drawings	25.00
566	1.19(b)(I Xi)	Certified copy of patent application as filed, regular service	15.00
567	1.19(b)(1)(ii)	Certified copy of patent application, expedited local service	30.00
568	1.19(b)(2)	Certified or uncertified copy of patent-related file wrapper and contents	150.00
569	1.19(b)(3)	Certified or uncertified copy of document, unless otherwise provided	25.00
570	1.19(b)(4)	For assignment records, abstract of title and certification, per patent	25.00
571	1.19(c)	Library service	50.00
572	1.19(d)	List of U.S. patents and SIRs in subclass	3.00
573	1.19(e)	Uncertified statement re status of maintenance fee payments	10.00
574	1.19(t)	Copy of non-U.S. document	25.00
575	1.19(g)	Comparing and certifying copies, per document, per copy	25.00
576	1.19(h)	Additional filing receipt, duplicate or corrected due to applicant error	25.00
577	1.21(c)	Disclosure document filing fee	10.00
578	1.21(d)	Local delivery box rental, per annum	50.00
579	1.21(e)	International type search report	40.00
580	1.21(g)	Self-service copy charge, per page	0.25
581	1.21(h)	Recording each patent assignment, agreement or other paper, per property	40.00
583	1.21(i)	Publication in *Official Gazette*	25.00
584	1.21(j)	Labor charges for services, per hour or fraction thereof	40.00
585	1.21(k)	Unspecified other services, excluding labor	AT COST
586	1.21(l)	Retaining abandoned application	130.00
587	1.21(n)	Handling fee for incomplete or improper application	130.00
588	1.21(o)	APS-Text terminal session time, per hour	40.00
592	1.21(k)	APS-CSIR terminal session time, per hour	50.00
590	1.24	Patent coupons	3.00
589	1.296	Handling fee for withdrawal of SIR	130.00
Patent	Enrollment Fees		
609	1.21(a)(1)(i)	Application fee (nonrefundable)	40.00

619	1.21(a)(1)(ii)	Registration examination fee	310.00
610	1.21(a)(2)	Registration to practice	100.00
611	1.21(a)(3)	Reinstatement to practice	40.00
612	1.21(a)(4)	Copy of certificate of good standing	10.00
613	1.21(a)(4)	Certificate of good standing – suitable for framing	20.00
615	1.21(a)(5)	Review of decision of Director, Office of Enrollment and Discipline	130.00
616	1.21(a)(6)(i)	Regrading of A.M. section (PTO Practice and Procedure)	230.00
620	1.21(a)(6)(ii)	Regrading of P.M. section (Claim Drafting)	230.00

		GENERAL FEES	
	Finance	Service Fees	
607	1.21(b)(1)	Establish deposit account	10.00
608	1.21(b)(2)	Service charge for below minimum balance	25.00
608	1.21(b)(3)	Service charge for below minimum balance restricted subscription deposit account	25.00
617	1.21(m)	Processing returned checks	50.00

Computer Service Fees

| 618 | Computer records | AT COST |

REMITTANCES FROM FOREIGN COUNTRIES MUST BE PAYABLE AND IMMEDIATELY
NEGOTIABLE IN THE UNITED STATES FOR THE FULL AMOUNT OF THE FEE REQUIRED

Code 37	CFR	Description	Fee Fee

Trademark Processing – Fees

361	2.6(a)(1)	Application for registration, per class	245.00
362	2.6(a)(2)	Filing an Amendment to Allege Use under ß I (c), per class	100.00
363	2.6(a)(3)	Filing a Statement of Use under ß I (d)(1), per class	100.00
364	2.6(a)(4)	Filing a Request for a Six-month Extension of Time for Filing a Statement of Use under ß I (d)(1), per class	100.00
365	2.6(a)(5)	Application for renewal, per	300.00
366	2.6(a)(6)	Additional fee for late renewal, per class	100.00
367	2.6(a)(7)	Publication of mark under ß 12(c), per class	100.00
368	2.6(a)(8)	Issuing new certificate of registration	100.00
369	2.6(a)(9)	Certificate of correction, registrant's error	100.00
370	2.6(a)(10)	Filing disclaimer to registration	100.00
371	2.6(a)(11)	Filing amendment to registration	100.00
372	2.6(a)(12)	Filing ß 8 affidavit, per class	100.00
373	2.6(a)(13)	Filing ß 1'5 affidavit, per class	100.00
374	2.6(a)(14)	Filing combined ßß 8 & 15 affidavit, per class	200.00
375	2.6(a)(15)	Petition to the commissioner	100.00
376	2.6(a)(16)	Petition for cancellation, per class	200.00
377	2.6(a)(17)	Notice of opposition, per class	200.00
378	2.6(a)(18)	Ex parte appeal, per class	100.00
379	2.6(a)(19)	Dividing an application, per new application (file wrapper) created	100.00

Trademark Service Fees

461	2.6(b)(1)(i)	Printed copy of each registered mark, regular service	3.00
462	2.6(b)(1)(ii)	Printed copy of each registered mark, overnight delivery to PTO box or overnight fax	6.00
463	2.6(b)(1)(iii)	Printed copy of each registered mark ordered via expedited mail or fax, exp. service	25.00
464	2.6(b)(4)(i)	Certified copy of registered mark, with title and/or status, regular service	15.00
465	2.6(b)(4)(ii)	Certified copy of registered mark, with title and/or status, expedited local service	30.00
466	2.6(b)(2)(i)	Certified copy of trademark application as filed regular service	15.00
467	2.6(b)(2)(ii)	Certified copy of trademark application as filed, expedited local service	30.00
468	2.6(b)(3)	Certified or uncertified copy of trademark-related file wrapper and contents	50.00
469	2.6(b)(5)	Certified or uncertified copy of trademark document, unless otherwise provided	25.00

470	2.6(b)(7)	For assignment records, abstracts of title and certification per registration	25.00
475	1.19(g)	Comparing and certifying copies, per document per copy	25.00
480	2.6(b)(9)	Self-service copy charge, per page	0.25
481	2.6(b)(6)	Recording trademark assignment agreement or other paper, first mark per document	40.00
482	2.6(b)(6)	For second and subsequent marks in the same document	25.00
484	2.6(b)(10)	Labor charges for services, per hour or fraction thereof	40.00
485	2.6(b)(11)	Unspecified other services, excluding labor	AT COST
488	2.6(b)(8)	X-SEARCH terminal session time, per hour	40.00
490	1.24	Trademark coupons	3.00

Fastener Quality Act Fees

650	2.7(a)	Recordal application fee	20.00
651	2.7(b)	Renewal application fee	20.00
652	2.7(c)	Late fee for renewal application	20.00

REMITTANCES FROM FOREIGN COUNTRIES MUST BE PAYABLE AND IMMEDIATELY NEGO-
TIABLE IN THE UNITED STATES FOR THE FULL AMOUNT OF THE FEE REQUIRED

APPENDIX VI

PTDLP Library Search Program

State	City/Library
Alabama	Auburn University: Ralph Brown Draughon Library* Birmingham Public Library
Alaska	Anchorage: Z.J. Loussac Public Library, Anchorage Municipal Libraries
Arizona	Tempe: Noble Science and Engineering Library, Arizona State University*
Arkansas	Little Rock: Arkansas State Library*
California	Los Angeles Public Library* Sacramento: California State Library San Diego Public Library San Francisco Public Library* Sunnyvale Center for Innovation, Invention & Ideas**
Colorado	Denver Public Library
Connecticut	Hartford Public Library New Haven Free Public Library

Delaware	Newark: University of Delaware Library
District of Columbia	Washington: Founders Library, Howard University
Florida	Fort Lauderdale: Broward County Main Library* Miami-Dade Public Library* Orlando: University of Central Florida Libraries Tampa Campus Library, University of South Florida
Georgia	Atlanta: Library & Information Center, Georgia Institute of Technology
Hawaii	Honolulu: Hawaii State Library*
Idaho	Moscow: University of Idaho Library
Illinois	Chicago Public Library Springfield: Illinois State Library
Indiana	Indianapolis-Marion County Public Library West Lafayette: Siegesmund Engineering Library, Purdue University
Iowa	Des Moines: State Library of Iowa
Kansas	Wichita: Ablah Library, Wichita State University*
Kentucky	Louisville Free Public Library*
Louisiana	Baton Rouge: Troy H. Middleton Library, Louisiana State University
Maine	Orono: Raymond H. Fogler Library, University of Maine
Maryland	College Park: Engineering and Physical Sciences Library, University of Maryland
Massachusetts	Amherst: Physical Sciences and Engineering Library, University of Massachusetts Boston Public Library*
Michigan	Ann Arbor: Media Union Library, The University of Michigan Big Rapids: Abigail S. Timme Library, Ferris State University Detroit: Great Lakes Patent and Trademark Center** Detroit Public Library

Minnesota	Minneapolis Public Library & Information Center*
Mississippi	Jackson: Mississippi Library Commission
Missouri	Kansas City: Linda Hall Library* St. Louis Public Library*
Montana	Butte: Montana Tech of the University of Montana Library
Nebraska	Lincoln: Engineering Library, University of Nebraska–Lincoln*
Nevada	Reno: University Library, University of Nevada–Reno
New Hampshire	Concord: New Hampshire State Library
New Jersey	Newark Public Library Piscataway: Library of Science and Medicine, Rutgers University
New Mexico	Albuquerque: Centennial Science and Engineering Library, The University of New Mexico
New York	Albany: New York State Library Buffalo and Erie County Public Library* New York: Science, Industry and Business Library, New York Public Library Stony Brook: Engineering Library, State University of New York
North Carolina	Raleigh: D.H. Hill Library, North Carolina State University*
North Dakota	Grand Forks: Chester Fritz Library, University of North Dakota
Ohio	Akron-Summit County Public Library Cincinnati: The Public Library of Cincinnati and Hamilton County Cleveland Public Library* Columbus: Ohio State University Libraries Toledo/Lucas County Public Library*
Oklahoma	Stillwater: Oklahoma State University*
Oregon	Portland: Lewis & Clark College
Pennsylvania	Philadelphia: The Free Library of* Pittsburgh The Carnegie Library of University Park: Pattee Library, Pennsylvania State University

Puerto Rico	Mayaguez: General Library, University of Puerto Rico
Rhode Island	Providence Public Library
South Carolina	Clemson: R.M. Cooper Library, Clemson University
Tennessee	Memphis & Shelby County Public Library & Information Center Nashville: Stevenson Science and Engineering Library, Vanderbilt University
Texas	Austin: McKinney Engineering Library, The University of Texas at Austin College Station: Sterling C. Evans Library, Texas A&M University* Dallas Public Library* Houston: The Fondren Library, Rice University** Lubbock: Texas Tech University Library
Utah	Salt Lake City: Marriott Library, University of Utah*
Vermont	Burlington: Bailey/Howe Library, University of Vermont
Virginia	Richmond: James Branch Cabell Library, VA Commonwealth University*
Washington	Seattle: Engineering Library, University of Washington*
West Virginia	Morgantown: Evansdale Library, West Virginia University*
Wisconsin	Madison: Kurt F. Wendt Library, University of Wisconsin–Madison Milwaukee Public Library
Wyoming	Casper: Natrona County Public Library

* Denotes APS-Text Access
** Denotes Partnership PTDL
PTDL program information is also found on the Internet at www.uspto.gov

APPENDIX VII

U.S. Copyright Application

FORM TX
For a Nondramatic Literary Work
UNITED STATES COPYRIGHT OFFICE

REGISTRATION NUMBER

_____

TX _____ TXU _____
EFFECTIVE DATE OF REGISTRATION

Month _____ Day _____ Year _____

DO NOT WRITE ABOVE THIS LINE. IF YOU NEED MORE SPACE, USE A SEPARATE CONTINUATION SHEET.

1

TITLE OF THIS WORK ▼

PREVIOUS OR ALTERNATIVE TITLES ▼

PUBLICATION AS A CONTRIBUTION If this work was published as a contribution to a periodical, serial, or collection, give information about the collective work in which the contribution appeared.   **Title of Collective Work ▼**

If published in a periodical or serial give: Volume ▼   Number ▼   Issue Date ▼   On Pages ▼

2 a

NAME OF AUTHOR ▼

DATES OF BIRTH AND DEATH
Year Born ▼   Year Died ▼

Was this contribution to the work a "work made for hire"?
☐ Yes
☐ No

AUTHOR'S NATIONALITY OR DOMICILE
Name of Country
OR { Citizen of ▶ _____
Domiciled in ▶ _____

WAS THIS AUTHOR'S CONTRIBUTION TO THE WORK
Anonymous?   ☐ Yes ☐ No
Pseudonymous?   ☐ Yes ☐ No
If the answer to either of these questions is "Yes," see detailed instructions.

NATURE OF AUTHORSHIP Briefly describe nature of material created by this author in which copyright is claimed. ▼

NOTE

Under the law, the "author" of a "work made for hire" is generally the employer, not the employee (see instructions). For any part of this work that was "made for hire" check "Yes" in the space provided, give the employer (or other person for whom the work was prepared) as "Author" of that part, and leave the space for dates of birth and death blank.

b

NAME OF AUTHOR ▼

DATES OF BIRTH AND DEATH
Year Born ▼   Year Died ▼

Was this contribution to the work a "work made for hire"?
☐ Yes
☐ No

AUTHOR'S NATIONALITY OR DOMICILE
Name of Country
OR { Citizen of ▶ _____
Domiciled in ▶ _____

WAS THIS AUTHOR'S CONTRIBUTION TO THE WORK
Anonymous?   ☐ Yes ☐ No
Pseudonymous?   ☐ Yes ☐ No
If the answer to either of these questions is "Yes," see detailed instructions.

NATURE OF AUTHORSHIP Briefly describe nature of material created by this author in which copyright is claimed. ▼

c

NAME OF AUTHOR ▼

DATES OF BIRTH AND DEATH
Year Born ▼   Year Died ▼

Was this contribution to the work a "work made for hire"?
☐ Yes
☐ No

AUTHOR'S NATIONALITY OR DOMICILE
Name of Country
OR { Citizen of ▶ _____
Domiciled in ▶ _____

WAS THIS AUTHOR'S CONTRIBUTION TO THE WORK
Anonymous?   ☐ Yes ☐ No
Pseudonymous?   ☐ Yes ☐ No
If the answer to either of these questions is "Yes," see detailed instructions.

NATURE OF AUTHORSHIP Briefly describe nature of material created by this author in which copyright is claimed. ▼

3 a

YEAR IN WHICH CREATION OF THIS WORK WAS COMPLETED This information must be given ◀ Year in all cases.

b DATE AND NATION OF FIRST PUBLICATION OF THIS PARTICULAR WORK Complete this information ONLY if this work has been published.   Month ▶ _____ Day ▶ _____ Year ▶ _____
◀ Nation

4

See instructions before completing this space.

COPYRIGHT CLAIMANT(S) Name and address must be given even if the claimant is the same as the author given in space 2. ▼

TRANSFER If the claimant(s) named here in space 4 is (are) different from the author(s) named in space 2, give a brief statement of how the claimant(s) obtained ownership of the copyright. ▼

APPLICATION RECEIVED
ONE DEPOSIT RECEIVED
TWO DEPOSITS RECEIVED
FUNDS RECEIVED

DO NOT WRITE HERE OFFICE USE ONLY

MORE ON BACK ▶ • Complete all applicable spaces (numbers 5-9) on the reverse side of this page.
• See detailed instructions.   • Sign the form at line 8.

DO NOT WRITE HERE
Page 1 of _____ pages

DO NOT WRITE ABOVE THIS LINE. IF YOU NEED MORE SPACE, USE A SEPARATE CONTINUATION SHEET.

PREVIOUS REGISTRATION Has registration for this work, or for an earlier version of this work, already been made in the Copyright Office?

☐ Yes ☐ No If your answer is "Yes," why is another registration being sought? (Check appropriate box) ▼

a. ☐ This is the first published edition of a work previously registered in unpublished form.

b. ☐ This is the first application submitted by this author as copyright claimant.

c. ☐ This is a changed version of the work, as shown by space 6 on this application.

If your answer is "Yes," give: **Previous Registration Number** ▼   **Year of Registration** ▼

5

DERIVATIVE WORK OR COMPILATION

a Preexisting Material Identify any preexisting work or works that this work is based on or incorporates. ▼

b Material Added to This Work Give a brief, general statement of the material that has been added to this work and in which copyright is claimed. ▼

6

See instructions before completing this space.

DEPOSIT ACCOUNT If the registration fee is to be charged to a Deposit Account established in the Copyright Office, give name and number of Account.

a Name ▼   Account Number ▼

CORRESPONDENCE Give name and address to which correspondence about this application should be sent.   Name/Address/Apt/City/State/ZIP ▼

b

Area code and daytime telephone number ▶   Fax number ▶

Email ▶

7

CERTIFICATION* I, the undersigned, hereby certify that I am the

Check only one ▶

☐ author
☐ other copyright claimant
☐ owner of exclusive right(s)
☐ authorized agent of _____

of the work identified in this application and that the statements made by me in this application are correct to the best of my knowledge.

Name of author or other copyright claimant, or owner of exclusive right(s) ▲

Typed or printed name and date ▼ If this application gives a date of publication in space 3, do not sign and submit it before that date.

Date ▶ _____

☛ Handwritten signature (X) ▼

X _

8

The filing fee of $20.00 is effective through December 31, 1998. After that date, please write the Copyright Office, check the Copyright Office Website at http://www.loc.gov/copyright, or call (202) 707-3000 for the latest fee information.

Mail certificate to:

Name ▼

Number/Street/Apt ▼

Certificate will be mailed in window envelope

City/State/ZIP ▼

YOU MUST:
• Complete all necessary spaces
• Sign your application in space 8

SEND ALL 3 ELEMENTS IN THE SAME PACKAGE:
1. Application form
2. Nonrefundable filing fee in check or money order payable to *Register of Copyrights*
3. Deposit material

MAIL TO:
Library of Congress
Copyright Office
101 Independence Avenue, S.E.
Washington, D.C. 20559-6000

9

APPENDIX VIII
CONTRACT FORMS:
Basic Boilerplate-Type
Agreements

PARTNERSHIP AGREEMENT

THIS AGREEMENT made and entered into at San Diego, California, this day of September, 1990 by and between _____and_____,
WITNESSETH:
The parties hereto agree as follows:

ARTICLE 1   FORMATION OF GENERAL PARTNERSHIP
Section 1.01 The parties hereto hereby form a general partnership.

ARTICLE 2   NAME AND PLACE OF BUSINESS
Section 2.01 The Partnership shall be conducted under the Partnership's name of _____ The principal place of business shall be of the Partnership shall be _____,_____

ARTICLE 3   PURPOSE
Section 3.01 The principal purpose of the Partnership is
_____

ARTICLE 4   TERM OF PARTNERSHIP

Section 4.01 The Partnership shall commence on the date hereof to continue until terminated by the agreement of a majority of the partners. Upon termination distribution shall be made as hereinafter provided.

ARTICLE 5   CAPITAL CONTRIBUTIONS
Section 5.01 a. shall each contribute his/her talents, efforts and energies to the business of the Partnership on a non-exclusive basis.

    The Partners shall each have a _____ percent interest in the Partnership; which interest may hereafter be adjusted by the partners' unanimous agreement of.

Section 5.02 From time to time the Partners may agree to contribute funds or property to the partnership. Such contributions shall constitute the Capital Contributions of the partner contributing same, and shall increase his/her interest as then agreed upon by a majority of the percentage of Partnership interest noted thereon by the partners.

Section 5.03 No Partner shall make an additional capital contribution unless agreed to by all the others.

ARTICLE 6   ACQUISITION OF PARTNERSHIP PROPERTY
Section 6.01 The Partnership shall acquire such properties and investments as the Partners shall agree upon. All and any acquisitions, contracts, promotions, franchises, joint participation or commission agreements and investment developments negotiated on behalf of the Partnership by any partner, shall be and become Partnership property.

ARTICLE 7   PROFIT AND LOSSES; CAPITAL
AND INCOME ACCOUNTS

Section 7.01 The net profits of the Partnership shall be divided and any losses shall be borne by the Partners in the percentages of their partnership interests.

Section 7.02 For purposes of this Agreement, the term "net profits of the Partnership" shall mean the net profits derived from operations of the business of the Partnership as ascertained in accordance with federal tax regulations as set out in the Internal Revenue Code and Regulations.

Section 7.03 An individual capital account shall be maintained for each Partner. The capital interest of each Partner in his capital account shall consist of his original contribution of capital, increased by additional capital contributions and by any amount transferred from his income account to his capital account pursuant to these Articles.

Section 7.04 An individual income account shall be maintained for each Partner. Each Partner's share of the net profits of the Partnership shall be credited to his income account, and each Partner's share of any net loss suffered by the Partnership shall be charged to his income account unless all the Partners agree to charge such loss to their capital accounts.

Section 7.05 A credit balance in a Partner's income account shall constitute a liability of the Partnership to that Partner, payable without interest, at such time as may be determined by vote of all the Partners, and shall not constitute a part of his interest in the capital of the Partnership. A debit balance in a Partner's income account, whether occasioned by withdrawals in excess of his share of the Partnership's net profits or the charging to his income account of his share of a Partnership net loss, shall constitute an obligation of such Partner to the Partnership, and shall reduce his interest in the capital of the Partnership to the extent thereof.

Section 7.06 For accounting and tax purposes only and notwithstanding any distributions made hereunder, the net income or losses, as ascertained through the use of the Internal Revenue Code and regulations promulgated thereunder, shall be allocated to each Partner in accordance with the percentage of his partnership interest.

ARTICLE 8 DISTRIBUTIONS
Section 8.01 No disbursements shall be made to any Partner unless the Partnership expenses are paid in full and all liabilities of the Partnership are then current; and then only with the express consent of the other Partners.

Section 8.02 If the Partnership has cash or other property which, in the unanimous, opinion of the Partners, is in excess of the needs of the Partnership, such may be distributed among the Partners in proportion to their interests in profits of the Partnership.

ARTICLE 9 BANKING

Section 9.01 The Partnership shall maintain a bank account in the name of the Partnership, with such bank the Partners shall determine, in which there shall be deposited all of the funds of the Partnership. No other funds shall be deposited in the account. The funds in aid account shall be used solely for the business of the Partnership, and all withdrawals therefrom are to be made by checks signed—by two of the Partners so authorized by all the Partners.

ARTICLE 10 BOOKS
Section 10.01 The Partnership shall maintain full and accurate books and

all Partners shall have the right to inspect and examine such books at reasonable times. Any Partner shall have the right to have the Partnership books audited. Within ninety (90) days after the expiration of each fiscal year, a balance sheet and a profit and loss statement prepared by the Partnership's accountant, together with a statement showing the capital accounts of each Partner, shall be distributed to each Partner and the amount thereof reportable for State and Federal income tax purposes.

ARTICLE 11   SALARIES, DRAWINGS, EXPENSE ACCOUNTS, AND INTEREST ON CAPITAL ACCOUNTS

Section 11.01 The Partners, and each of them, shall not receive compensation for services rendered to and on behalf of the Partnership.

Section 11.02 No Partner shall receive any interest on his contribution to the capital of the Partnership. Any advance of money to the Partnership by any Partner, made with the consent of the Partners, in excess of such Partner's agreed capital contribution, shall not be deemed a capital contribution to the Partnership but a debt due from the Partnership to such Partner and shall be repaid within thirty (30) days, with interest at such rates and times as determined by all the Partners. Provided, however, that any Partner may make an additional contribution of capital to the Partnership if, all, the other Partners agree in writing that such advance of funds or other contributions shall be a contribution of capital to the Partnership.

ARTICLE 12   MANAGEMENT, DUTIES, POWERS AND RESTRICTIONS

Section 12.01 The Partners shall devote only so much of their time and attention to the Partnership business as he shall deem reasonable or necessary for the furtherance of such business; provided, however, all of the Partners may designate a Managing Partner who shall be primarily responsible for the management of the business of the Partnership. Such managing partner shall receive compensation for his services as is agreed upon by all the Partners.

Section 12.02 Each Partner shall have a voice in the management and conduct of the Partnership business pro rata to his interest in the Partnership. Any differences arising among the Partners as to ordinary matters connected with the Partnership business shall be cause for dissolution, unless resolved by unanimous agreement.

Section 12.03 No Partner shall incur any obligation on behalf of the Partnership exceeding the sum of ($1, 000.00 without the consent of all the other Partners. Any Partner who incurs any such obligation in violation of this provision shall be individually liable to the other Partners therefore.

Section 12.04 The Partnership shall indemnify a Partner for payments made and personal liabilities reasonably incurred by him in the ordinary and proper conduct of the Partnership business, or for the preservation of the Partnership business or property.

Section 12.05 No Partner shall, without the consent of all other Partners:
a) Loan any Partnership funds;
b) Extend Partnership credit to any person a Partner has notified the Partnership not to trust;
c) Incur any obligations in the name or on the credit of the Partnership, except in the ordinary course of the Partnership business;
d) Become bailee, surety, or endorser for any other person;
e) Do any act which would make it impossible to carry on the ordinary
f) business of the Partnership;
g) Confess a judgment against the Partnership;
h) Admit to the Partnership another general partner. Any loss sustained by the Partnership because of the breach of this paragraph by any Partner shall be deducted from such or, if the net profits of the Partnership for the fiscal year in which the breach occurred be insufficient, from such Partner's capital interest in the Partnership.

Section 12.06 Each Partner shall. protect the Partnership from all costs, claims, and demands in relation to his individual obligations.
Section 12.07 No Partner shall sell, assign, mortgage, hypothecate, or encumber his interest in the Partnership. Further, no Partner shall make any assignment of his Partnership interest for the benefit of his creditors, or transfer such interest to a trustee or receiver for the benefit of his creditors.

ARTICLE 13   ACTS OF DISSOLUTION AND WINDING UP OF PARTNERSHIP

Section 13.01 Insolvency. In the event that a majority in interest of the Partners should elect to cause the dissolution and winding up of the Partnership in accordance with Section 13.07, then same shall be so dissolved.

Section 13.02 Voluntary Withdrawal. Any partner may voluntarily withdraw from the Partnership either by selling his interest in the Partnership; however, in such event he shall first offer same to the remaining partners at the same price, terms and conditions; and in such event the remaining partners may purchase same in the same proportions as their partnership interests.

Section 13.03 Purchase by Less than All Partners. Should any one or more of the remaining Partners be unable or unwilling to exercise the option to purchase the interest of a withdrawing Partner, such option may be exercised and such interest purchased by the other remaining Partners.

Section 13.04 Assumption of Obligation. On any purchase and sale of a Partnership interest being made as in this Article provided, the remaining Partners shall assume all the Partnership obligations and shall protect and indemnify the withdrawing Partner from liability for any such obligations.

Section 13.05 Notice of Dissolution. On any purchase and sale of a Partnership interest being made as in this Article provided, the remaining Partners shall, at their own cost and expense, as soon as reasonably practicable after exercise of their option to purchase, or as soon as reasonably practicable after the purchase and sale of such interest, cause to be prepared, published, filed and served all such notices as may be required by law to protect the withdrawing Partner from liability for future obligations of the Partnership business.

Section 13.06 Special Provision in the Event of Death. On dissolution of the Partnership by reason of the death of a Partner, the remaining Partners may continue the business of the Partnership to the end of the calendar month in which such death occurs, and the estate of the deceased Partner shall share in the net profits and losses of the Partnership during the balance of such month in the same manner as the deceased Partner would have had he survived to the end of such month. Provided, however, the estate of any such deceased Partner shall not be liable for any obligations or losses incurred by the Partnership after the date of such deceased Partner's death beyond the value of the deceased Partner's interest in the Partnership at the date of his death.

Section 13.07 Winding Up. On dissolution of the Partnership, the affairs of the Partnership shall be wound up, the assets liquidated, the debts paid, and the remaining funds divided among the Partners according to their then Partnership interests.

ARTICLE 14   MISCELLANEOUS
Section 14.01 Any and all notices between the parties hereto provided for or permitted under these Articles or by law shall be in writing and shall be deemed duly served when personally delivered to a Partner or, in lieu of such personal service, when deposited in the United States mail, certified, postage prepaid, addressed to such Partner at his address.

Section 14.02 Any and all consents and agreements provided for or permitted by these Articles shall be in writing, and a signed copy thereof shall be filed and kept with the books of the Partnership.

Section 14.03 Should any litigation be commenced between the parties hereto or their personal representatives concerning any provision of these Articles or the rights and duties of any person in relation thereto, the party or parties prevailing in such litigation shall be entitled, in addition to such other relief as may be granted, to a reasonable sum as and for his or their attorney's fees in such litigation which shall be determined by the court in such litigation or in a separate action brought for that purpose.

Section 14.04 This instrument contains the sole and only agreement of the parties hereto relating to their Partnership and correctly sets forth the rights, duties and obligations of each to the others as of its date. Any prior agreements, promises, negotiations, or representations not expressly set forth in these Articles are of no force and effect.

Section 14.05 Unless named in this Agreement or unless admitted to - the Partnership as' hereinabove provided, no person shall be considered a Partner; and the Partnership, each, Partner and any other persons having business with the Partnership need deal only with the Partners so named or so admitted; and they shall not be required to deal with any other person by reason of an assignment or transfer by a Partner or by reason of the death of a Partner, except as otherwise provided in this Agreement. (in accordance with this Agreement, in the absence of a transfer of the legal ownership of the Partnership interest of a transferring or deceased Partner, any payment by the Partnership to the person shown on the Partnership records as a Partner, or to his legal representatives, or to the assignee of the right to receive Partnership distributions, shall acquit the Partnership of all liability to any other person who may be interested in such payment by reason of an assignment by the Partner, or by reason of his/her death, or for any other reason.

IN WITNESS WHEREOF the parties have executed this Agreement the date above written.

_____

_____

INDEPENDENT DISTRIBUTOR AGREEMENT

This Agreement is entered into on the date set forth herein by and between _____, Incorporated, a Delaware Corporation with its corporate offices located in the State of _____ (the "Producer" or the "Company") and the undersigned Distributor, (the "Distributor").

      1. <u>Background</u>. Producer formulates and packages certain technological cosmetic formulations currently packaged and sold under the trade name of "_____" (tm) (the "Product"), and desires to create a nationwide network of Independent Distribution Centers ("IDC's") to sell and distribute the Product to retail accounts and also directly to the consumer. Producer desires to increase the public's awareness and consumption of the Product and to prevent detrimental, injurious and uneconomic practices in the distribution of said Product. Distributor desires to secure an exclusive geographical area within which to sell said Product.

      2. <u>License</u>. Subject to those limitations set forth herein and in subsequent written communications to Distributor from Producer, and provided that all Product advertisements created by Distributor must have Producer's prior written approval, Producer hereby grants to Distributor the exclusive and nontransferable right and license to:

(i) market and sell the Product at wholesale prices to retail outlets within that exclusive geographic area defined in Exhibit A (the "Territory"); and

(ii) network and sell the product at retail prices directly to the consumer with no geographical boundaries; and

(iii) use Producer's trademarks, labels, copyrights, and other Company produced advertising commercials in the marketing of the Product within the Territory.

      3. <u>Covenants of Producer.</u> Producer covenants as follows:

      (i) so long as Distributor shall not at any time be in default herewith, Producer shall supply Distributor with the Product in the regular course of business under terms and conditions no less favorable dm those granted to any other Distributor to whom Producer sells similar quantities of the Product. Such terms, conditions and quantities may from time to time be modified or changed in Producer's sole discretion;

(ii) Producer shall ship Product in a good and marketable condition with a "shelf life" consistent within the trade for similar products;

(iii) Producer shall make available for Distributor to purchase various brochures, literature, displays, samples and other promotional material at prices listed in Producer's then current promotional material price list;

(iv) Producer shall be responsible for the quality of the Product while in its care, custody or control. This responsibility, however, shifts to the Distributor upon delivery of the Product to Distributor's shipper at Producer's warehouse. Producer will supply Distributor with any special instructions necessary to store and maintain the Product and shall label the Product with instructions to the consumer as to use and storage. Distributor shall return any defective product and notify Producer within ninety (90) days after delivery if any such Product proves to be defective, whereupon Producer shall replace the same without further cost to Distributor so long as Distributor has notified Producer within ninety (90) days and returned said Product within ninety (90) days;

(v) Producer will promptly send Distributor all inquiries received from any potential customers regarding purchases of Product in the Territory which inquiries would inure to the benefit of Distributor;

(vi) So long as Distributor demonstrates that it is not in any manner responsible for the alteration or damage of the Product its container, its labels, or any promotional or advertising material of Producer, nor made any claim or representation pertaining to the Product other than as set forth in Producer's literature, Producer will indemnify and hold the Distributor harmless from any damages, legal costs or other expense reasonably resulting from any judgment, claim or defense of any claim against Distributor resulting from an defective Product;

(vii) PROVIDED, HOWEVER Producer shall not be in default hereunder should its failure to perform pertain to or arise out of any cause beyond its control, including but not limited to, acts of God, fires, floods, strikes, freight, embargoes or acts of third parties or of governmental agencies, or otherwise; and

(viii) Producer shall have the exclusive right, in its sole discretion, to add, delete or modify its product line.

4.   Covenants of Distributor. Distributor covenants as follows:

(i) Along with this signed Agreement, Distributor shall submit an initial payment to the Producer as described in

Exhibit A hereto;

(ii) Without Producer's prior written permission, Distributor shall not sell or be directly or indirectly involved in marketing, merchandising, promoting or handling in any way merchandise which is competitive with any products manufactured, distributed or sold by Producer;

(iii) Distributor shall be responsible for and shall promptly pay all excise, use or sales taxes pertaining to the Product or its sale required to be paid by any governmental or taxing agency;

(iv) Distributor shall maintain the Product under reasonable warehouse conditions as specified from time to time by Producer. Distributor will be solely responsible for any Product mishandled or improperly stored following its delivery by Producer;

(v) Distributor agrees that it is an independent contractor, is not for any purpose an employee or agent of Producer, and any personal employed by Distributor shall solely be employees or agents of Distributor and not of Producer,

(vi) Without the prior written permission of the Producer, Distributor shall at no time have the authority, expressed or implied, to sign contracts or other obligations, nor to make purchases, nor to acquire or dispose of any property other than the product for or on behalf of the Producer,

(vii) Distributor shall post in a conspicuous area of the IDC a sign, window decal or other marketing of sufficient size and print reasonably calculated to indicate to and inform the public that the IDC is operated by an independent contractor and not by Producer. Distributor shall not use stationery, invoices, checks, advertisements or other method, instrument or device which does not clearly reflect Distributor's status as an independent contractor;

(viii) Distributor hereby agrees to indemnify and hold Producer harmless from all expenses, costs, including but not limited to attorneys' fees, claim and liabilities of any land whatsoever, whether by operation of law or otherwise, arising out of either: a) Distributors failure to inform any third-party of Distributors lack of authority to bind the Producer for any purpose, or b) any Product altered or damaged by or at the behest of

Distributor or its agents, or c) any Product as to which Distributor has made any claim or representation or advertising not previously approved by Producer;

(ix) Distributor agrees that any of Producer's proprietary confidential information disclosed to or discovered by Distributor resulting from the relationship created by this Agreement shall be held in confidence, used only for the performance of this Agreement, and in all respects remain the sole property of the Producer;

(x) Distributor shall promptly reform Producer in writing of any infringement or Violation of Producer's trademarks, tradenames, labels and/or copyrights.

5. **Proprietary Rights**. This Agreement does not confer upon Distributor any proprietary right, title or interest in the trade names, trademarks, menus, formulae, trade secrets or other property of Producer.

6. **Cost of Shipping**. Distributor shall purchase the Product "F.O.B." Producer's warehouse, with all costs of shipping, insurance, and related transportation costs to be borne by Distributor.

7. **Price Changes**. Producer reserves the right at any time to change the price, and agrees to promptly inform Distributor thereof in writing, in which event Distributor may cancel its future obligations hereunder by written notice thereof delivered to Producer within twenty (20) days of Producer's notice, in which event (Distributor's choice to cancel this Agreement) Producer shall have no duty to repurchase any of Distributor's unused inventory nor advertising materials.

8. **Terms of Payment**. Distributor shall pre-pay Producer's invoice for all Product orders until Distributor establishes credit according to the Producer's "Standard Credit Policy to Distributors" and credit is authorized in writing by Producer.

9. **Assignment**. Any attempted assignment of its duties or benefits hereunder by Distributor without the prior written consent of Producer shall be void and of no legal effect and shall constitute Distributor's default hereunder. Producer's approval of assignment will not be unreasonably withheld, but may be conditioned upon Producer's prior satisfaction of the following:

A) The potential transferee has an acceptable credit rating, is of good moral character, and possesses business and Professional qualifications satisfactory to Producer;

B) Distributor shall not have been in default of any tern, condition, or covenant of this agreement and shall have fully Paid and satisfied all accrued obligations hereunder;

C) The proposed transferee must execute a new IDC agreement and any other agreements that the Producer may, at the time of the transfer, be requiring of new licensees;

D) The proposed transferee satisfies the requirements of the Producer's Training Program; and

E) Distributor has at that time Purchased the minimum inventory then required by Producer for new IDC'S.

10. **Exclusive Sales Territory**. Distributor shall be the exclusive distributor of the Product within that geographic area described in Exhibit A hereto SUBJECT To Distributor maintaining that Minimum Order Level described in Exhibit A hereto. This Minimum Order Level is known as the "IDC Minimum", and is subject to change from time to time by Producer and is subject to Producer giving Distributor at least ninety (90 days written notice of said change, in which event Distributor may cancel its future obligations hereunder by written notice thereof delivered to Producer within twenty (20) days of Producer's notice, in which event (Distributor's choice to cancel Agreement) Producer shall have no duty to repurchase any of Distributor's unused inventory or advertising material.

11. **Advertising Claims**. Distributor may advertise the Product in good taste, and in a truthful and non-misleading manner, after obtaining Producer's prior written approval of all proposed advertising, which approval shall not be unreasonably withheld. Distributor shall not in any manner employ any misleading or deceptive claim pertaining to the Product Producer, or Distributor.

12. **Endorsements**. Distributor may use any Producer's current endorsers in appropriate and truthful advertisements after first obtaining Producer's current certification that said endorser(s) are at such time under contract with Producer.

13. **Other Products**. Producer may from time to time produce other product(s) other than the Product described here and Producer will offer said product(s) to the Distributor and Distributor is under no obligation to carry said product(s).

14. **Termination**.

(i) Distributor may cancel this Agreement without the necessity for cause upon ninety (90) days written notice to the Producer so long as Distributor has paid in full all outstanding obligations to Producer;

(ii) Producer, at its sole discretion, may terminate this Agreement at any time should the Distributor commit any criminal acts, fraud, unethical practices, breach Territory lines, or default hereunder;

(iii) If the Producer becomes party to an acquisition, merger, initial public offering or the sale of a majority of its assets, then the new Producer entity has thirty days to provide written notice of termination or re-negotiation of this Agreement or this Agreement shall remain in full force and effect; if this Agreement is terminated per this clause (iii) then the Distributor must be reimbursed for any unsold inventory and unused promotional material.

(iv) Should any voluntary or involuntary petition in Bankruptcy be filed on behalf of Distributor or should Distributor make an assignment of the Product for the benefit of creditors, then, at Producer's sole discretion, any obligations of Producer to Distributor shall terminate; and

(v) In the event of any termination described in Paragraph 14, Producer shall have no duty to repurchase unused inventory nor advertising materials nor to refund any portion of the territory fee except as provided in subparagraph (iii).

15. **Arbitration**. All disputes between the parties hereto arising out of this Agreement shall be submitted by the parties to binding arbitration in Boston, Massachusetts pursuant to the provisions of the Massachusetts Civil Code. Both parties hereby agree to be bound by the decision of said Arbitration. The prevailing party shall recover its legal fees and expenses. Any award may be reduced to a stipulated judgment which, if not immediately paid, may be recorded in all appropriate jurisdictions.

16. **Effective Date**. This agreement is effective on the date set forth in Paragraph 19 and will remain in full force and effect or unless otherwise canceled for cause, per Section 14, or by legal proceeding.

17. **Entire Agreement**. This Agreement constitutes the entire understanding of the parties, supersedes any prior oral or written agreement or understanding, may be amended only by writings and shall be

construed in accordance with the laws of the State of _____.
(This space deliberately left blank.)

18. **Notice**. Notice hereunder shall be by personal service, Federal Express, or certified U.S. mail, return receipt requested postage pre-paid, to the following addresses. It shall be the responsibility of the party whose address is changed to so notify the other party.

Producer: **Distributor:**
XYZ, Incorporated   _____

_____

_____

19.   Agreement Date. Ibis Agreement is entered into this, the

_____
day of _____, 19

DISTRIBUTOR: PRODUCER:

XYZ, Incorporated
a Delaware Corporation

EXHIBIT A

INDEPENDENT DISTRIBUTOR
CENTERS EXCLUSIVE TERRITORY

MINIMUM ORDER LEVEL, AND INITIAL PAYMENT

1. **The Territory**. Subject to the provisions of the attached "Independent Distributor Agreement," the Distributor therein named pertaining to the Product therein defined shall have the exclusive right to market and sell the Product at wholesale prices to retail outlets in that sales territory (the "Territory") comprising that geographic area of the United States of America encompassing the following zip code area(s):

2. **Minimum Order Level for Territory**. The distributor shall order from the Producer a minimum of _____volume dollars of Product per quarter year which minimum shall be known as the "IDC Minimum." The IDC Minimum shall commence 60 days from the date of this agreement.

Initial Payment & Product Order. The Distributor's initial payment and product order to Producer, to be paid simultaneously with the delivery of the attached signed "Independent Distributor Agreement" is as follows:

a. Initial Product Order (Territory Fee) $_____
b. Advertising Materials $_____

TOTAL $_____

TRADEMARK LICENSE AGREEMENT

MEMORANDUM OF AGREEMENT, made this 17 day of December 1990, by and between_____, INC., a corporation orga-nized under the laws of the state of_____, United States of America, having a principal place of business in_____, _____ (hereinafter "_____") and _____, an individual having a principal place of business at _____ (hereinafter "licensee").

WITNESSETH

WHEREAS, DCSI is a corporation owned and controlled by _____ with the authority to market, license and promote the use of_____name, trademarks, trade dress or brands as shown on the annexed Schedule A, alone and in connection with the campaign _____.

WHEREAS, Licensee is desirous of obtaining exclusive license to market, promote, use and sell various products and/or conduct various services which bear the _____ name, trademarks, brand dress or brand, along or in connection with the campaign for _____, within a specific territory under the terms and conditions herein set forth; and WHEREAS, DCSI is willing to grant such exclusive license to Licensee. NOW, THEREFORE, in consideration of the mutual undertak-ings herein set forth and other good and valuable consideration, receipt of which is acknowledged, the parties agree as follows:

1.   **DEFINITIONS**
        1.1 "Assigned Territory" shall mean United States and Canada.
        1.2 "Trademarks" shall mean all statutory and common law trademark rights, registered and unregistered, vested in DCSI in the Assigned Territory in connection with trademarks, tradenames, trade dress, service marks or brands as set forth in Schedule A.
        1.3 "Licensed Products" shall refer to the products or services to be marketed by Licensee under this Agreement, namely _____.

2.   **GRANT OF LICENSE**
        2.1 DCSI hereby grants to Licensee for the term of the License an exclusive non-transferable license within the Assigned Territory to use the Trademarks in connection with the marketing and sale of its Licensed Products.

3.   **TERM**
        3.1 This License Agreement shall be effective as of the date first above written and continue in force and effect for a term of two years thereafter (the "Term").

3.2 Upon expiration of the Term, Licensee will undertake, with due dispatch, to cease all further use of the Trademarks on or in connection with its products. Within a reasonable period of time, not to exceed six (6) months following the date of such expiration hereof, all use of the Trademarks shall cease.

4.   PAYMENT TO DCSI

4.1. In consideration for the foregoing grant, Licensee agrees to make payments to DCSI as follows:

4.1.1 A licensing fee to be paid in accordance with Schedule B attached hereto.

4.1.2 A royalty calculated as a percentage of net sales, as Shown in Schedule C attached hereto, to be paid monthly. "Net Sales" shall mean the aggregation of all amounts invoiced to the purchasers of the Licensed Products sold by the Licensee, (i) not including costs of freight and shipping insurance charges and taxes or duties which may be included in the invoice and (ii) less volume discounts. Royalties shall accrue in favor of DCSI upon the shipment of the Licensed Products by Licensee.

5.   RECORDS AND REPORTS

5.1 The Licensee agrees, during the term of the License, to keep adequate and complete records and all information necessary to verify the sales of Licensed Products and the time and amount of royalties to DCSI under Section 4.1.2 hereof. These records shall be open to inspection by DCSI or its authorized representatives during reasonable business hours to the extent necessary to make such verification. On the 15th day of each month, Licensee shall also render to DCSI an accounting of all of the sales of Licensed Products made during the immediately preceding month. The accounting shall indicate the sales made, the net selling prices, deductions, credits and allowance, etc., thus making it clear to DCSI how its accrued royalties, if any, were determined. Such an accounting shall be rendered whether or not any money is due and payable. Licensee shall forward to DCSI concurrently with such accounting any royalty payment that accrued in favor of DCSI during the immediately preceding month.

6.   GOODWILL

6.1 Licensee expressly acknowledges the value of DCSI's goodwill in the Trademarks and further expressly agrees that any goodwill generated as a result of this License shall inure to the benefit of DCSI.

7.   QUALITY CONTROL

7.1 Licensee shall agree to manufacture Licensed products bearing the Trademarks to satisfy DCSI's standards of quality, in order to preserve and not denigrate, the goodwill symbolized by the Trademarks.

7.2 Licensee warrants that the licensed Products will be of good quality in design, material, and workmanship and will be suitable for their intended purpose; that no injurious, deleterious, or toxic substances will be used in or on the Licensed Products; that the Licensed Products will not cause harm when used as instructed and with ordinary care for their intended purpose; and that Licensed Products will be manufactured, sold and distributed in strict compliance with all applicable laws and regulations.

7.3 Licensee agrees that DCSI shall have the right to approve or disapprove in advance of sale the quality, style, design, colors, appearance, materials and workmanship of all Licensed Products and the components thereof, and to approve or disapprove any and all Trademarks used on or in connection with the Licensed Products. Licensee shall not distribute or sell any such Licensed Product which has not been approved by DCSI or which is, at any time, disapproved by DCSI.

8.   COVENANT AGAINST CHALLENGE

8.1 Licensee is expressly prohibited from challenging or contesting, in any way, the validity or enforceability of the Trademarks which are the subject of this License Agreement or ownership thereof by DCSI.

9.   COVENANTS

9.1 Licensee is expressly prohibited from any sublicensing under this Agreement without the express written consent of the DCSI.

9.2 Licensee is expressly prohibited from assigning this agreement or any rights hereunder, in whole or in part, without the express written consent of DCSI.

9.3 Licensee agrees to diligently market and sell the Licensed products to the widest extent that its capabilities permit. If, within three (3) months following the execution of this agreement, Licensee has failed to take good faith steps to exploit the rights granted to it hereunder, DCSI shall have the right to terminate this agreement immediately by giving Licensee written notice of termination. Licensee agrees to manufacture the Licensed Products in sufficient quantities to meet the reasonably anticipated demand. Licensee also agrees to exercise reasonable efforts to advertise and promote the Licensed Products at its own expense and to use its best efforts to sell the Licensed Products in the Assigned Territory.

9.4 Licensee shall not engage in any marketing practice involving the Licensed Products to which practice DCSI disapproves; however, DCSI is in no way suggesting or mandating any particular marketing method, plan or system.

10. REPRESENTATIONS AND INDEMNIFICATIONS

10.1 DCSI represents that to the best of its knowledge it has good title to the Trademarks and is unaware of any other party's rights which might be infringed by the use of the Trademarks in connection with the Licensed Products.

10.2 DCSI makes no other representations or warranties of any kind whatsoever, except as expressly set forth above. In this connection, and without limiting the generality of the foregoing, DCSI states that it makes no representations or warranties regarding the scope of protection, validity or enforceability of the Trademarks when used on the Licensed Products, or the commercial efficiency or value of the Trademarks.

10.3 Licensee hereby indemnities DCSI with respect to any and all claims, asserted liabilities, costs, reasonable attorneys' fees, judgments or damages arising from Licensee's production and/or sale of the Licensed Products and Licensee's performance under this License Agreement.

11. SUPPRESSION OF INFRINGEMENT

11.1 In the event of any infringement of the Trademarks DCSI shall have the sole right at its election to suppress such infringement by litigation or otherwise, which suppression shall be controlled by DCSI at its sole expense. In the event DCSI elects not to suppress such infringement, then, in that event only, Licensee shall have the right to pursue such suppression at its sole expense. Under any circumstances, Licensee shall immediately notify DCSI of any potential trademark infringements which come to its attention.

12. RIGHT OF FIRST REFUSAL

12.1 For a period of one year following the termination of this agreement, DCSI shall not grant the right to use the Trademarks in the manufacture or marketing of Licensed Products without first offering such right to licensee under terms and conditions, that are substantially similar to the proposed third party offer. Such offer shall remain outstanding for five business days. If written notice of Licensee's acceptance of such offer is not received by DCSI within those five business days, DCSI shall be free to enter into proposed transaction with the third party.

13. TERMINATION

13.1 Notwithstanding the foregoing, this license agreement may be terminated by DCSI under any of the following circumstances:

13.1.1 Licensee breaches a material provision of this license agreement unless within thirty (30) days after written notice of the breach licensee corrects the same;

13.1.2 Licensee becomes insolvent or commits an act of bankruptcy;

13.1.3 Licensee fails to maintain the normal conduct of its business, determined by the nature and level of its business activities and operation during the calendar quarter immediately preceding the date on which DCSI contends the normal business conduct was not maintained.

14. APPLICABLE LAW AND FORUM SELECTION

14.1 The parties agree that all matters of dispute that arise at any time by reason of any of the terms of this License Agreement of the validity or enforceability of the Trademarks or any registration thereon shall be litigated in an appropriate court having jurisdiction in the Southern District of California, United States of America, and the laws of the State of California shall govern in all respects without reference to conflict-of -laws principles, and Licensee consents to jurisdiction in said forum.

15. NOTICES AND COMMUNICATIONS

15.1 All notices and communications shall be in writing and shall be mailed and delivered to the business addresses of the respective parties:

To: _____

_____

With a copy to:
To Licensee: _____

_____

16. INVALIDITY

16.1 If any of the provisions of this Agreement are deemed to be invalid or unenforceable, the invalidity of such provision shall not affect the validity or enforceability of any other provision.

17. BINDING EFFECT

17.1 This Agreement shall be binding upon and inure to the benefit of the parties and their respective successors, assigns and heirs.

18. WAIVER

18.1 A waiver by either party of any breach of any provisions of this Agreement shall not operate or be construed as a waiver of any other breach or a subsequent breach of the same or a different provision.

19. Complimentary product during the term of this agreement, licensee will provide to DCSI, at no charge, the following; 100 each _____ .

20. ENTIRE UNDERSTANDING

20.1 This Agreement constitutes the entire understanding between the parties with respect to the subject matter hereof. No modification, extension or waiver of any provision hereof or any release of any rights hereunder shall be valid unless the same is in writing and is consented to by both parties hereto.

21. ENTRIES AND DELETIONS

The parties hereby affirm that all entries and deletions made in the blanks throughout this Memorandum of Agreement, whether handwritten or typed, are correct and have been entered with their consent and understanding.

IN WITNESS WHEREOF, the parties hereto have executed this Agreement as of the date above written.

_____,Inc.

Dated:_____, 2___
By:_____

Dated:_____, 2___
By:_____

(SIGNATURE PAGE TO TRADEMARK LICENSE AGREEMENT)

APPENDIX IX

DOD SBIR Grant Mailing List Request

The DOD SBIR Program Office maintains a computerized listing of firms that have requested to be sent copies of the DOD SBIR Solicitations on a regular basis. If you would like to be, remain, or be added to this listing, please mail in this form.

❏ YES, include my name and address on the DOD SBIR Mailing List
❏ NO, remove my name and address from the DOD SBIR Mailing List

NAME: _____

COMPANY: _____

ADDRESS: _____

CITY: _____ STATE: _____ ZIP: _____

PHONE: _____(___)_____

To send: Remove this page, fold along the marked lines on the reverse side, seal with tape or staple, and affix postage.

Is this a new address?   ☐ YES   ☐ NO

Old Address: _____

_____

_____